THE HUMAN FOOD CHAIN

Proceedings of the Conference 'The Food Chain: Forging The Links', organised by the Faculty of Agriculture and Food, University of Reading, and held at the University of Reading, 29–31 March, 1988.

The Conference represented a collaborative effort between Departments of the Faculty. Four Working Groups each studied separate aspects of the Food Chain for about 18 months prior to the Conference. Their final conclusions emerged as the four Themes which comprised the Conference itself.

The four Chairmen of the Working Groups were:

Professor M. I. Gurr

Dr J. G. W. Jones

Professor H. E. Nursten

Professor J. S. Marsh

who, together with Professor C. R. W. Spedding, Dr G. G. Birch (who served as co-ordinator) and J. A. Burns, formed the Organising Committee, chaired initially by C. R. W. Spedding and subsequently by J. S. Marsh.

The editing of the proceedings therefore involved all of the foregoing and C. R. W. Spedding, who initiated the enterprise, was asked to take the responsibility of being Editor on behalf of all those concerned.

THE HUMAN FOOD CHAIN

Edited by

C. R. W. SPEDDING

*Professor of Agricultural Systems and Director of the Centre for
Agricultural Strategy, University of Reading, UK*

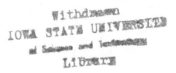

ELSEVIER APPLIED SCIENCE
LONDON and NEW YORK

ELSEVIER SCIENCE PUBLISHERS LTD
Crown House, Linton Road, Barking, Essex IG11 8JU, England

Sole Distributor in the USA and Canada
ELSEVIER SCIENCE PUBLISHING CO., INC.
655 Avenue of the Americas, New York, NY 10010, USA

WITH 41 TABLES AND 10 ILLUSTRATIONS

© 1989 ELSEVIER SCIENCE PUBLISHERS LTD

British Library Cataloguing in Publication Data

Food Chain: Forging the Links
    The human food chain: (proceedings of the conference The Human Food Chain:
    Forging the Links held at the University of Reading, 29–31 March 1988)
    Food Production Industries
    I. Title.   II. Spedding, C. W. R. (Colin Raymond William)
    338.1.'9

Library of Congress Cataloging in Publication Data

The human food chain/edited by C.R.W. Spedding.
        p.   cm.
    Proceedings of a conference entitled 'The food chain: forging the
links.' held at the University of Reading. March 29–31. 1988.
    Bibliography: p.
    Includes index.
    ISBN 1–85166–317–7
    1. Food industry and trade — Congresses.   2. Agriculture —
Congresses.   3. Food — Research — Congresses.   4. Agricultural
education — Congresses.   5. Food service — Study and teaching —
Congresses.   6. Cookery — Energy consumption — Congresses.
I. Spedding C. R. W.
HD9000.5.H8   1989   88-33704
363.8 — dc 19

Phototypesetting by Tech-Set. Gateshead, Tyne & Wear.
Printed in Great Britain at the University Press, Cambridge.

Acknowledgements

The Organising Committee wishes to acknowledge financial support
for this Conference from:

ICI plc Plant Protection Division
Cadbury Schweppes plc
Safeway Food Stores Limited
Expoconsult
Tate and Lyle plc
Dalgety Agriculture Limited
Chalcombe Publications

The Committee is most grateful for this support, without which the
important preparatory work for the Conference could not have been
undertaken.

Contents

THEME 3: IMPLICATIONS FOR TECHNOLOGY:
PRIORITIES FOR R & D

List of Contributors

V. D. ARTHEY

Campden Food and Drink Research Association, Chipping Campden, Gloucestershire, GL55 6LD, UK.

MARGARET ASHWELL

The British Nutrition Foundation, 15 Belgrave Square, London, SW1X 8PS, UK.

J. A. BURNS

Department of Agricultural Economics and Management, University of Reading, No. 4 Earley Gate, Reading, RG6 2AR, UK.

GEOFFREY CAMPBELL-PLATT

Department of Food Science and Technology, University of Reading, Whiteknights, PO Box 226, Reading, RG6 2AP, UK.

PETRONELLA CLARKE

Nutrition Unit, Department of Health, Alexander Fleming House, Elephant and Castle, London, SE1 6BY, UK.

A. CROSSLEY

Unilever Research Laboratory, Colworth House, Sharnbrook, Bedford, MK44 1LQ, UK.

J. B. DENT

Edinburgh School of Agriculture, University of Edinburgh, West Mains Road, Edinburgh, EH9 3JG, UK.

A. M. FRECKLETON

Unilever Research, Port Sunlight Laboratory, Bebington, Merseyside, L63 3JW, UK.
Present address: Paragon Communications Limited, Film House, 142 Wardour Street, London, W1V 3AU.

W. GRANT

Department of Politics, University of Warwick, Coventry, CV4 7AL, UK.

M. I. GURR

Milk Marketing Board, Thames Ditton, Surrey, KT7 0EL, UK.

FRANK HARDING

Milk Marketing Board, Thames Ditton, Surrey, KT7 0EL, UK.

G. HARRINGTON

Planning and Development, Meat and Livestock Commission, PO Box 44, Queensway House, Bletchley, Milton Keynes, MK2 2EF, UK.

SIMON HARRIS

British Sugar plc, PO Box 26, Oundle Road, Peterborough, PE2 9QU, UK.

J. G. W. JONES

Department of Agriculture, University of Reading, No. 1 Earley Gate, Reading, RG6 2AT, UK.

M. G. LINDLEY

Tate & Lyle plc, Group R & D, Whiteknights, Reading, RG6 2BX, UK.

MAUREEN MALIK

Department of Hotel and Catering Studies, Huddersfield Polytechnic, Queensgate, Huddersfield, HD1 3DH, UK.

J. S. MARSH

Department of Agricultural Economics and Management, University of Reading, No. 4 Earley Gate, Reading, RG6 2AR, UK.

A. D. McCLUMPHA

The Nestlé Company Limited, St. Georges House, Croydon, Surrey, CR9 1NR, UK.

D. McHALE

Cadbury Schweppes plc, The Lord Zuckerman Research Centre, University of Reading, Whiteknights, Reading, RG6 2BX, UK.

H. E. NURSTEN

>   *Department of Food Science and Technology, University of Reading,*
>   *Whiteknights, PO Box 226, Reading, RG6 2AP, UK.*

D. P. RICHARDSON

>   *The Nestlé Company Limited, St. Georges House, Croydon, Surrey,*
>   *CR9 1NR, UK.*

B. A. ROLLS

>   *AFRC Institute of Food Research, Shinfield, Reading, RG2 9AT, UK.*

J. D. SCHOFIELD

>   *Flour Milling and Baking Research Association, Chorleywood,*
>   *Hertfordshire, WD3 5SH, UK.*

C. R. W. SPEDDING

>   *Department of Agriculture, University of Reading, No. 1 Earley Gate,*
>   *Reading, RG6 2AT, UK.*

T. E. SWINDELL

>   *Glaxo Group Research, Greenford Road, Greenford, Middlesex,*
>   *UB6 0HE, UK.*

R. THOMAS

>   *Bonair, 2 The Avenue, Clevedon, Avon, BS21 7ED, UK.*

SHEILA A. TURNER

Department of Science Education, University of London Institute of Education, 20 Bedford Way, London, WC1H 0AL, UK.

A. F. WALKER

Department of Food Science and Technology, University of Reading, Whiteknights, PO Box 226, Reading, RG6 2AP, UK.

ANN WEST

Department of Hotel and Catering Studies, Huddersfield Polytechnic, Queensgate, Huddersfield, HD1 3DH, UK.

A. J. YOUNGS

Department of Education and Science, Ringway House, Ringway, Preston, PR1 3HQ, Lancashire, UK.

C. A. ZAROR

Department of Food Science and Technology, University of Reading, Whiteknights, PO Box 226, Reading, RG6 2AP, UK.

The Food Chain: Forging the Links

C. R. W. SPEDDING

Department of Agriculture, University of Reading, UK

INTRODUCTION

The industry as a whole is concerned with the food chain, from those who provide inputs to the primary producer to those who retail what the manufacturer processes and, indeed, beyond this, up to the final consumer.

Although each part of the industry may deal with only one sector, it is increasingly necessary for each to have regard to what is happening in or what is required by the rest of the food chain.

In particular, the interface between agriculture and food is becoming of increasing importance.

This is true of both developed and developing countries, although the situation is entirely different in each. In developing countries, much of the processing and preparation of food is done by the farmer, so the production and utilisation of food is often well integrated. What is often missing is integration with the supply of fuel needed to cook the food. Where food is traded, storage and transport to the consumer are often extremely inefficient and wasteful.

In developed countries, food production and food processing are mostly done by different people; indeed, by different industries. This separation has disadvantages and there needs to be a much greater awareness, by each sector, of the activities and objectives of the other.

In arriving at research priorities within agriculture, for example, it is necessary to take account of the needs of the food industry. Similarly, when considering food research it is essential to take account of what is possible in agriculture. The food industry has to say what it wants from agriculture and from food science and technology research, and also,

1

what kind of people it needs with what sort of qualifications and experience.

Agriculture, in both its research and educational programmes, needs to recognise the fact that up to 80% of our food is now processed in one way or another: the food industry is thus the major customer for agricultural output.

The ultimate consumer, however, includes all of us and we all feel entitled to judge what we like to eat, even if not always what is good for us. Consumer preferences therefore matter and a study of the whole food chain must involve social science research as well as that in the natural sciences. Indeed, there is probably no more multidisciplinary subject than the food chain and it provides — or should provide — a meeting point for many disciplines in an area that is central to all our lives.

Inevitably, the agriculture/food industry will continue to need specialists in all the important aspects of production and processing, but, in addition, there may be a need for people with a much broader grasp of the way the whole industry works. The question for educational establishments is how best to produce such people.

The same applies to research. Specialised research will continue to be needed but there is also a need for multidisciplinary research at the interfaces. The question here is how best to ensure and encourage the necessary collaborative research between those mainly concerned with agriculture or food.

The importance of a whole view of the food chain can be illustrated with reference to the use of support energy.

SUPPORT ENERGY USE IN THE FOOD CHAIN

Support energy includes all those sources of energy other than solar radiation and it is used to a considerable extent in the agricultural industries of all developed countries (see Table 1 for examples).

The sources of support energy are mainly fossil fuels and they are used not only to run tractors but also to make them; not only to apply fertiliser but also to manufacture it, and other inputs.

The absolute amounts of support energy are small compared with the solar radiation used by crops and even smaller in comparison to that actually received. Every hectare of land receives about 33 million MJ per year, of which about 825 000 MJ are stored in a wheat crop, for example,

TABLE 1
Proportion of Total Support Energy used by Agriculture and
Food in Developed Countries

Country	%	Reference
Agriculture:		
UK (1984)	1	1
USA (1980)	3	2
Europe (1980)	3–4	2
UK (1973)	3·9	3
Australia (1975)	2	5
Food system:		
USA (1973–6)	14–16	2
UK (1973)	15·8	3
USA (1977)	10·9	4

whereas some 11 000 MJ of support energy are used in its production.

If the efficiency of support energy use is calculated for wheat production, it gives a ratio of about 3 · 2, representing the MJ of energy in the product per MJ of support energy used to produce it.

Of course, none of the support energy used actually appears in the wheat: what is found in the crop is derived from solar radiation and photosynthesis. Nevertheless, this is how the calculation is normally made. If the same calculation is made for milk production, the ratio is about 0·6. It thus appears to be a very much less efficient process than wheat production and this is no surprise since all animal production processes are less efficient than crop production (shown as support energy cost of production in Table 2). This is mainly because animals have to live on crop products and thus add further inefficiencies of conversion on top of those involved in growing the crop. Indeed, the comparison with milk minimises the contrast, since milk production is the most efficient of the animal production processes carried out on farms.

Thus we may conclude that, if support energy became either scarce or costly, it would be sensible to move towards crop production and away from animal production.

But these calculations only apply up to the farm gate and yet 70–80% of our food is now processed before consumption (Table 3), most of it incurs packaging, storage and distribution costs, and most of it is cooked.

TABLE 2
Support Energy (SE) use in Crop and Animal Production[6]

Product	MJ of SE used per MJ energy produced
Barley	0·4
Maize	0·4
Winter wheat	0·3
Potatoes	0·6
Milk	1·85–2·8
Hen eggs	7·6
Broiler poultry	10·0
Lamb	4·3
Beef	9·4

TABLE 3
Agriculture's Production of Raw Materials for the Food (and Beverage) Industry[7]

Farm gate product	% Processed[a]	Examples of product
Raw cows' milk	97[b]	Pasteurised or sterilised liquid milk, yoghurt, cheese and butter
Sugar beet roots	100	Sugar, molasses and pulp
Cereals	100	Flour, breakfast food and malt
Live animals for slaughter	100	Carcase meat, cooked, cured and smoked meats, pies
Raw potatoes	51[b] (The remainder will be cooked before consumption)	Canned, crisped, dehydrated and frozen

[a]In addition to transporting and marketing.
[b]1979/80.

TABLE 4
Energy used (MJ) for Processing, Packaging and Distribution[8, 9]

Process	1 kg loaf		Beef (140 g) USA
	UK	USA	
Production on farm	4·02	13·8	121·3
Processing	15·99	8·2	0·2
Packaging, etc.	0·71	8·6	1·8
Total	20·72	30·6	123·3

Energy analysis up to the farm gate is thus a very partial calculation and may give a quite misleading impression. As it happens, in the example used, milk can be drunk virtually as it is and usually incurs only minor additional energy costs in transport and treatment.

Wheat, by contrast, is never consumed whole and is normally ground and cooked, involving processes that incur very heavy energy costs, quite apart from the costs of storage, transport and packaging (see Table 4).

These comparisons for milk and wheat are shown in Table 5, where it can be seen that, at the point of consumption, there is very little difference in the support energy ratios and, indeed, in this particular calculation, milk has the higher value.

It is obvious that, for this to happen, the efficiency of support energy use during the period from farm gate to consumption must be reversed: that is, it must be much higher for milk. The conclusion from looking at the food industry would thus be the opposite of that derived from looking at the production processes, in response to increased scarcity and cost of support energy.

TABLE 5
Efficiency of Energy Use[8, 10]

Product	MJ in product per MJ SE used
Wheat at farm gate	3·2
Bread — white, sliced, wrapped	0·5
Milk at farm gate	0·65
Milk bottled and delivered	0·595

TABLE 6
Efficiency of Support Energy Use at Different Points in the Food Chain[8, 10]

Product	Farm gate	Farm gate to consumer	Total
Milk	0·284	3·176	0·261
Wheat/bread	7·429	0·635	0·558
Sugar beet/refined sugar	3·589	6·041	2·251

$$\text{Efficiency} = \frac{\text{Gross energy in product}}{\text{SE used in production}}$$

Three totally different conclusions are reached, therefore, by looking (a) up to the farm gate (i.e. at agriculture), (b) from the farm gate to the consumer (i.e. at the food industry) and (c) at the whole process (i.e. at the food chain). Any number of such examples could be given and they do not all come out the same way, depending upon the amount and nature of the processing carried out. Table 6 gives selected calculations, all based on energy production, but similar results can also be obtained for protein production.

TABLE 7
SE as % of Total Costs[11]

System	Support energy
Dairy	
Intensive	77·8
Extensive	69·9
Beef	
Intensive	87·9
Extensive	74·0
Sows	
Intensive	79·2
Extensive	70·5
Eggs	
Intensive	81·4
Extensive	75·4
Broilers	82·6
Pigs	86·4

The reason for choosing support energy as the input is (a) that its price and availability does vary greatly and it depends upon non-renewable resources, and (b) support energy costs, as a proportion of total costs, are quite high — up to 80% for intensive meat production but still very important in extensive enterprises (Table 7).

This focus on energy is by no means the only one worth pursuing but it has been recognised as of such importance that a whole international organisation (The Agri-Energy Roundtable) was set up in 1980 to draw attention to the interlinked futures of world food and energy production and their effect on international economic and political relations. Its approach is a unique attempt to bring together all those involved in 'food systems', including representatives of all sectors of the food chain from 'field to fork'.

However, this is only one illustration of the need for the food and farm industries to develop stronger interactions. Another is the increasing role of farming as a supplier of raw materials to industry — and not only the food industry.

AGRICULTURE AS A SUPPLIER OF RAW MATERIALS

As shown in Table 3, the major purchaser of farm produce is the food industry. Those engaged in farming, therefore, need to know more of the way in which the food industry works, what it can do and what it needs or can use. Similarly, especially in relation to new products, those in the food industry need to know more about what farming can and does produce and its potential for future development.

This is particularly so because the ultimate customer may often be operating a negative preference. A recent example has been the objection of some customers to eggs and meat produced in batteries. This is based on considerations of animal welfare. Yet it does not follow that eggs produced in other ways (e.g. aviaries or even 'free range') have necessarily been 'better' for the hen. Battery cages actually protect the hen from predators, pests, parasites, diseases and the weather: they are not simply driven by direct economic factors. Alternative systems may put one thing right, only to fall foul of another.

What is required in such cases is an innovative effort by *all* concerned to meet the needs of the customer, where the customer only knows what he/she is against. This is not uncommon in relation to animal production: nor can the customer be expected to know what is a better system.

C. R. W. Spedding

FOOD SAFETY

Similar difficulties occur in relation to *food safety*. Everyone is concerned that food should be safe for the consumer but perceptions as to both the problems and the solutions vary greatly.

The public tends to focus on such questions as 'additives', whilst the food industry would put a higher priority on the avoidance of bacterial contamination. But the public is also concerned about the use of agrochemicals during production, including the effects on the animal and on the environment, and about the relationships between diet and health.

Food safety is thus a subject which permeates the entire food chain and needs to be considered in this light.

There are, however, some examples where the customer has very definite views as to what he or she wishes to have available. Organically-produced food is one such example.

ORGANIC FARMING

Although 'organic' can legitimately be applied to all kinds of farming, it has come to mean a particular way of producing food, typically without added chemicals as pesticides, herbicides or fertilisers. But, in fact, it usually involves much more than this, including rotational practices and particular ways of keeping animals.

The fact is that there is an increasing number of people, in the UK and in the rest of Europe, who wish, for a variety of reasons, to purchase food produced in this fashion. But, once produced organically, neither the producers nor prospective purchasers want the products 'spoiled' by additives or other practices they dislike occurring during processing, packaging or distribution. Thus we have a group of consumers *and* producers who are vitally concerned with the whole food chain and may even want to keep it completely separate from any other food chain. So products, such as 'organic' or 'free-range', may need to be labelled, not only with regard to content and composition, but also to be identifiable in terms of the food chain that produced them.

Increasingly, therefore, in the development of new products, whether crop- or animal-based, it is going to be necessary for all parts of the food chain to work together. This applies to those engaged in R & D, as already illustrated, and both have to reflect what is wanted and feasible at both ends of the food chain.

THE NEED FOR CHANGE

But such changes cannot easily occur, when those trained and educated — and those doing the training and educating — are largely separated from each other.

This argument also has powerful implications for agricultural and food policymaking and the organisational arrangements that underlie these processes. In the UK, we have one ministry responsible for both agriculture and food, so it is at least possible for an integrated policy to be developed. But research and education tend to remain separate in these two areas.

A massive change in outlook is required, not least in education, and at Reading, where we have a large Faculty of Agriculture and Food, we feel a responsibility to explore what should be done.

The first step is to encourage new thinking about the issue and that is what this conference is designed to do.

REFERENCES

1. Department of Energy (1985) *Digest of UK Energy Statistics 1985*, HMSO, London.
2. Fluck, R. C. and Baird, C. D. (1980) *Agricultural Energetics*, AVI, Westport, Connecticut.
3. White, D. J. (1977) *Phil. Trans. R. Soc.*, **281**, 261–75.
4. Marsh, J. S. (1980) *Span*, **23**(3), 102–4.
5. Millington, R. J. (1975) *Wld Anim. Rev.*, **16**, 18–22.
6. Spedding, C. R. W. (1982) In: *Energy Management and Agriculture*, D. W. Robinson and R. C. Mollan (Eds), Proc. 1st Int. Summer School in Agriculture, Dublin, 337–49.
7. Spedding, C. R. W. (1985) In: *The Future Demands Change*, Proc. 39th Oxford Farming Conference, 99–113.
8. Leach, G. (1976) *Energy and Food Production*, IPC Science & Technology Press.
9. Pimentel, D. and Pimentel, M. (1979) *Food, Energy and Society*, Edward Arnold, London.
10. Walsingham, J. M. (1984) Personal communication.
11. Spedding, C. R. W., Thompson, A. M. M. and Jones, M. R. (1983) *Agro-Ecosystems*, **8**, 169–81.

THEME 1

Public Perception and Understanding

Introduction

M. I. GURR

Milk Marketing Board, Thames Ditton, Surrey, UK

In this theme, we shall be concerned with what the various people involved in the food chain understand about the different stages in the chain and how information is communicated. We shall be specially concerned with consumers since it is the satisfaction of consumers' needs for food that is the whole purpose of the chain. We do, however, attach great importance to the notion that the producer should understand the problems and motivations of the processor and retailer and vice versa. Nevertheless, most of the investigations that have generated quantitative or qualitative information have been concerned with messages directed at or coming from consumers and we make no apology for simplifying our discussion by concentrating on consumer aspects. Furthermore, we will make an additional simplification by emphasizing one particular kind of message, namely those concerned with nutrition. This, in part, derives from the special expertise of our sub-committee and partly from the high level of media attention given to this subject recently and its resulting influence on all parts of the food chain. Other sorts of messages may be equally important, for example, concerns about the contamination of foods by pesticides, radioactivity, heavy metals or microorganisms; the employment of chemicals in farming or in food processing and the exploitation of animals in agriculture. They will be alluded to where appropriate but the theme of nutrition in relation to health and disease will be used as an example of the way in which better communication is needed between participants in the food chain.

This section of the conference addresses issues of what might be termed 'non-formal education' and relates to all other sections but particularly that on formal education. We address three main themes:

The medium — how do people acquire their nutrition information?
The message — what sort of nutrition information is being, or should be relayed and to whom? and
The motivation — how can we encourage people to want to become better educated about nutrition and about the food chain in general? How can we encourage scientists, teachers and the media to interact better to convey that information?

In the first part, we probe the whole question of information and misinformation in nutrition. Misinformation can arise at any point in the information chain and may stem from misunderstanding of the data or the desire of individuals to achieve some sort of prominence, the area of their endeavours being pure chance. The best defence is a public sufficiently educated and sophisticated to recognize nonsense when they see it. If this is so, a heavy responsibility is placed on education, both formal and 'popular'.

The second part examines the types of nutrition messages in circulation, their origins and to whom they are directed. We conclude that government, the media, the nutrition and medical professions, food manufacturers and multiple retailers all have roles in conveying such information and improving education but their activities need co-ordinating and monitoring.

Part three discusses the involvement of food manufacturers and retailers in conveying nutrition messages through various types of claims about their products. This leads to part four in which we explore in detail a nutrition labelling system designed to inform, educate and motivate.

We conclude that more thought needs to be given to the language of communication. Terms that are readily understood by professionals ('saturated fat', 'energy', 'fatty acid' for example) are known to be at best incomprehensible and at worst, downright misleading to many consumers and we need to take greater care in the presentation of information in lay terms.

We also need to give greater attention to the targets of our educational programmes. Each of us is a consumer and each has a different background, degree of knowledge and education, and differing needs. Attitudes are strongly determined by upbringing: the more ingrained, the more difficult are they to change. Perhaps we should be taking the long term view and concentrating our resources only on the very young who are tomorrow's food purchasers. Initiatives in nutrition education can only be successful in the context of the education system as a whole. More emphasis on the scientific method and on the practical value of science in the community is needed. Only then can the broadening of syllabuses to include nutrition and health as a component of the schools' curricula be effective.

Given good early nutrition education, the way is paved for consumer education through a coordinated programme of promotion and labelling. Although the final expression of this information will be in the hands of the manufacturers and retailers, it will require a network of support involving government, agriculturalists and the nutrition and medical professions. In short, this means no less than the whole-hearted support of all elements of the food chain — a theme that is at the heart of this conference.

The proceedings of this section conclude with a summary of the discussion, which although not revealing any significant new concepts, usefully extended the thinking on the three major themes. An appendix summarizes the results of a survey, undertaken during the course of the working group's deliberations, detailing sources of nutrition information used by the media and providing insights into editors' views of nutrition journalism. The detailed results will be published elsewhere.

Public Perception and Understanding

A. M. FRECKLETON[a]*, M. I. GURR[b], D. P. RICHARDSON[c], B. A. ROLLS[d] and
A. F. WALKER[e]

[a]Unilever Research, Port Sunlight Laboratory, Merseyside, UK
[b]Milk Marketing Board, Thames Ditton, Surrey, UK
[c]The Nestlé Company Limited, Croydon, Surrey, UK
[d]AFRC Institute of Food Research, Reading, UK
[e]Department of Food Science and Technology, University of Reading, UK

ABSTRACT

Although the link between diet and 'the diseases of affluence' is not proven and degenerative diseases are regarded as having complex multifactorial aetiologies, many national and international expert committees have recommended dietary changes in an attempt to reduce their incidence. In the UK, dietary guidelines such as the COMA (1984) report represent a change from past nutritional advice which was concerned with avoiding deficiency diseases. Now the emphasis is on the balance of macronutrients in the diet. COMA recommended that 'those responsible for health education should inform the general public of the recommendations and how to implement them. In particular, advice should be given on how to construct diets and regulate physical activity in order to minimise the risk of cardiovascular disease and avoid obesity'.

This report mainly addresses issues of non-formal education. Although there is evidence that simple nutritional messages delivered by the media may be assimilated quickly by the public, fundamental education is slow and difficult. Thus, knowledge of nutrition is held at various levels, with those groups who would benefit most being hardest to reach. Even so, knowledge of

*Present address: Paragon Communications Limited, London.

17

nutrition does not necessarily mean that the people of any group will change their eating habits, but it is a necessary acquisition on the way to such change. The effectiveness of education initiatives can be studied at three levels by monitoring changes: (a) in consumer knowledge of nutrition, (b) in eating patterns and (c) in morbidity and mortality statistics. There are some published data on (a) and (b) from consumer surveys and these are dealt with later in this report and, although it is early days since the publication of the UK guidelines, a DHSS/MAFF survey will provide data on (c).

INTRODUCTION: UNDERSTANDING THE FOOD CHAIN

What do we understand by the food chain and who is concerned in it? Figure 1 indicates the various stages involved in getting food from primary producers to the consumers, the many influences on these stages and the people operating them. In this paper we are going to be concerned with what the people involved in the food chain understand about the different stages. We shall be especially concerned with

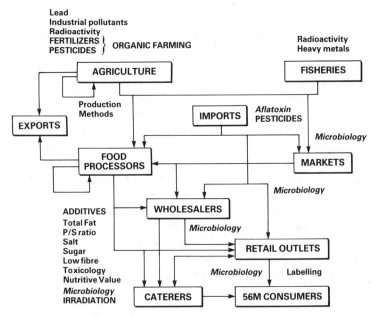

FIG. 1. The food chain. Bold lower case, general concerns; capitals, mainly consumer-led concerns; italics, mainly science-led concerns.

consumers since it is the satisfaction of the consumers' needs for food that is the whole purpose of the chain. We do, however, attach importance to the notion that the producer should understand the problems and the motivation of the processor and vice versa.

Nevertheless, most investigations which have generated qualitative or quantitative information have been concerned with messages directed at or coming from consumers and our discussion will be simplified by concentrating on this aspect. It will be further simplified by emphasizing one particular kind of message, namely nutritional ones, because of the predominant expertise of our working group and of the high level of media attention given to this subject recently. Other sorts of messages (indicated in Fig. 1) may be equally or even more important, for example concerns about contamination of foods by pesticides, radioactivity, heavy metals or microorganisms; the employment of chemicals in farming or in food processing and the exploitation of animals in agriculture. They will be alluded to where appropriate but the theme of nutrition in relation to health and disease will be used as an example of the way in which better communication is needed between participants in the food chain, particularly consumers.

INFORMATION AND MISINFORMATION IN NUTRITION: THE MEDIUM, THE MESSAGE AND THE MOTIVATION

The Medium

How do people acquire their nutrition information? We can be fairly confident that few will consult the scientific literature directly; this is not to ascribe intellectual laziness to people, for each of us is an uninformed layperson in many areas. Since no-one has the time to become an expert in every field, we must accept that primary information is filtered, processed and promulgated by intermediates.

A few will be advised directly by a health professional, a dietition or general practitioner. This usually happens, however, when disease is already manifest: few physicians have the time to advise the general patient (and may feel themselves inadequately informed); dietitians normally treat only referred patients. A number of publications, of variable quality, emanate from government departments or committees. There are some outstanding examples, but many steer an uneasy compromise between misleading by oversimplification and alienating the general reader who is felt to be unequipped or unwilling to understand and assess technical detail. Some, of course, are aimed largely at the professional, or to advise members of the food industry on legal requirements.

Additional sources of nutrition information are the various components of the food chain. Manufacturers may issue advertising material highlighting the contribution of their products to the well-being of consumers; producers or retailers of certain products may produce supporting material, especially if they feel their food(s) to be unfairly assailed as harmful. One assumes — perhaps optimistically — that they will not be so unwise as actually to lie, but in the nature of things, flattering evidence will be stressed while negative research will be omitted or de-emphasized. Large retailers have begun to give their customers nutrition information, some of good quality.

By far the greatest volume of nutrition information (and misinformation) comes, however, from the 'popular' media: radio; television; newspapers, daily and weekly, national and local, quality and tabloid; women's and family magazines; specialized 'health' or dietary magazines; popular nutrition books of all kinds, many of which enjoy large sales. Many of these are written by people with no formal training in nutrition or life science, and some by those with apparently no knowledge or understanding of the subject. It is not surprising that some strange views are put forward and, presumably, believed.

A critical question that arises is: how are these spokespeople selected and what checks are made on them and their material? This question is one that we hope to address in our postal questionnaire (see Appendix). One obvious difficulty is that although degrees and diplomas in nutrition are conferred by respectable academic institutions, there is — at present — nothing to prevent an individual lacking any training or qualification being represented as a nutrition expert.

Finally, few people receive any instruction in nutrition at their primary or secondary schools. This is a subject to be addressed in another theme.

The Message

At what stage does misinformation enter the public domain? There are several sources and one must remain aware of two kinds of error in each of them. Some emanate from sincere people who show every sign of believing their message; others from those whose behaviour is cynical and manipulative. It is difficult to know where to draw the line and, although it is almost impossible to believe that some material is intended seriously, each person's assessment will depend on his or her capacity for crediting human stupidity or villainy.

One source of error that we must not forget is that the best information available to us may be uncertain or just plain wrong, so that a proper consensus may be impossible or, in the light of fuller knowledge, misleading. (One should not forget that, a relatively short time ago, those suffering from pernicious anaemia had all their teeth extracted to promote a 'cure'.) Health issues are often complex, and to deliberate misrepresentation and unjustified extrapolation must be added 'ignorant misrepresentation' — a failure to grasp the information. The cautious COMA[1] recommendations (that it seems on balance that certain dietary changes will result in a fall in the incidence of cardiovascular disease, although the information falls short of proof) are interpreted to mean that dietary reform will eliminate heart disease from the community. Many writers confuse (or have, perhaps, never understood) the distinction between the population and the individual. (Certain actions may reduce one's risk of disease; it is still possible to do all the 'right' things and suffer a fatal coronary at forty.) A good example of how readily statistics may be misunderstood has been pointed out by Wheelock.[3] The meaning read into a report by Doll and Peto by the London Food Commission shows that they have not only misinterpreted the conclusion but have no grasp of epidemiology or statistics.

The current tendency to equate 'natural' with 'good' in every circumstance demonstrates how an intuitive feeling can become a matter of faith. A brief survey of the known facts — or even a few moments thought — would serve to show that whether or not a food constituent exists naturally is quite irrelevant to whether or not it is harmful.

There seems to be a deeply held belief that the general public, those to whom politicians patronizingly refer as 'ordinary people', should not have information laid before them and invited to decide; instead they should be instructed, frightened, bullied or hectored into adopting an approved lifestyle. The writers of popular nutrition articles, especially those of an evangelical persuasion, will extirpate the caveats and reservations in the original material to produce a direct, simple message pruned of inconvenient detail and evidence inconsistent with the point being made. Thus may the reader be presented with a set of instructions uncomplicated by information that may arouse doubt. It seems to be a sound general rule that the more dogmatic the pronouncement, the more tenuous is the writer's grasp of the strengths and limitations of experimental science. It is not only the advertiser who replaces 'may' or 'can', under some circumstances by 'will' and for editorial material there is no equivalent of the Advertising Standards Council.

Only the most innocent and unworldly can believe that every newspaper exists to give a fair and balanced account of important matters so that the reader may reach informed conclusions. Newspapers — particularly at the popular end of the market — exist to sell newspapers, and what sells newspapers is (or is felt to be, which may be as important) a mixture of entertainment and flattery. If the truth is uninteresting, a feature will simply be invented.[4] Nutritionists would be naive to expect that their own field should be treated any differently. Dramatic headlines are *de rigueur*, preferably with a hint of scandal in high places. Thus, we have meaningless lists of 'good and bad' foods, advice that would be impractical for the population as a whole (game should replace beef and lamb), statements that are demonstrably wrong (our diet is the single largest cause of ill health), lists of MPs who back 'good food' (presumably the rest wish to poison the electorate) and, perhaps the most breathtaking, the list of poisonous food additives that editors revive from time to time, possibly when they are bored. If a sizeable section — or even a vociferous section — of the readership is convinced of a 'fact', few, if any, editors will consider it their function to suggest that they may be in error.

Tempting as is the target that popular journalism presents, active workers in nutrition should be aware that the impression they give may not be one that, on reflection, they would wish to have given. A conference of cardiovascular surgeons is unlikely to play down the significance of cholesterol; an individual who has spent his or her life studying iron absorption may not point out that anaemia is relatively rare in the UK population. It is important that scientists in general, and nutritionists in particular, give more thought to their own skills in communication.

The Motivation

Since nutrition is a multidisciplinary subject, some of whose fields are less well developed than others, we are not in a position to prescribe, with absolute confidence, the diet and lifestyle that would unfailingly promote a long and healthy life for each individual. Certainly, there is sufficient evidence to make recommendations that, on balance, may well be beneficial and should certainly do no harm. A lifestyle that includes moderate exercise, a prudent but not extreme diet, the avoidance of obesity and excess and (very importantly) a wise choice of parents will probably lead to a greater sense of well-being and may well delay the onset of degenerative disease.

Why should people who have as much access to information as anyone else put forward extreme and dogmatic versions of tentative suggestions? Why, moreover, should they attribute anything less than wholehearted endorsement of their views to a conspiracy deriving from industrial greed and speculation, bureaucratic incompetence and inaction. The fact that the food on our tables is at the end of a food chain resistant to overnight change is considered an irrelevance. Those who pointed out difficulties or entertain reservations are treated, not as sincere individuals who have reached different conclusions but as heretics — villains in the pay of multinational corporations or buffoons unable to grasp self-evident fact.

We must admit that there is just enough evidence of officialdom that is uninterested rather than disinterested and of corporations foolish enough to use unscrupulous methods to suppress unwelcome evidence to lend some credence to this picture. (Although the 'stupidity theory' is at least as plausible as the 'conspiracy theory'.) But the critical word is 'heretics': activists see food policy as a moral issue and adherence or non-adherence to a viewpoint as a question of faith. We would like to suggest that this is because the food issue draws on two basic strands of human motivation: fear of extinction and fear of (or resentment of) personal insignificance.

If a prescribed lifestyle can be promoted as postponing death or incapacity, the likelihood is that it will be embraced enthusiastically and uncritically. Preference is not for detailed discussion of the multifactorial influences on the development of a disease, but for a simple formula for 'salvation' — health, longevity, prolonged youth, increased sexual vigour. It has often been said that no one has gone out of business by underestimating public taste. Similarly, if cynically, it can be suggested that no one has failed to achieve prominence by underestimating public willingness to follow logical argument and assess evidence critically. Just as the religious person exalts faith above reason, the nutritional evangelist expounding a theory is essentially explaining himself or herself; thus contradiction will be met with anger and abuse.

As soon as this basic postulate is accepted, previously puzzling manifestations become explicable. The natural human preference for the prepackaged theory that makes it unnecessary for him to think can be seen in the dismissive attitude towards unwelcome facts. There seems to be little doubt that of the six sources of food-borne hazard, microbiological contamination represents the greatest danger, whereas food additives represent the least risk.[5] Yet the complexities of microbiology are largely ignored while article after article depicts additives as known poisons added quite unnecessarily to the diet by ruthless food companies for their own convenience. Such organizations are caught in an almost archetypal example of Morton's Fork: if they admit that some additives may affect some people adversely, the matter is proved; if they deny it, they are lying. Some manufacturers and processors, reasoning that if a reasonable segment of consumers want additive-free foods, they can have them, regardless of whether or not they are superior, rush to claim their market share. This is taken as further proof (if it were needed) that the omitted additives were harmful all along. Since (for example) additive-free bread becomes mouldy surprisingly quickly and must be discarded and replaced, increased sales may result.

Perhaps those who accede to such demands are wise. Spokespeople who try to suggest that hypersensitivity is a rare phenomenon are accused of implying that the customer rather than the food, is at fault. (It may be too obvious to spell out, but no one suggests that the allergic individual is morally at fault; this is an example of attack by unwarranted extension, not to mention a failure to comprehend the meaning of moral responsibility.)

Another frequently seen manifestation of the belief that opponents must necessarily be scoundrels is the assumption that all work funded by special interest groups is automatically suspect. A survey showing that (say) people eating two or more apples a day suffer fewer heart attacks would be accepted if funded by the MRC but regarded with intense suspicion if supported by the Granny Smith Development Board. The reports may be identical, but suggestions that all work should be judged on its merits are received with incredulity. For some reason, claims made for 'health' foods are exempt from suspicion: the most imaginative claims based on sketchy anecdotal evidence are believed uncritically. Accepting a salary or research funding from a commercial source is taken as proof of gross moral turpitude; to derive an income by writing scaremongering, inaccurate and internally self-contradictory articles is apparently perfectly acceptable. This is not to deny that in a perfect world legislators and commentators would be able to consult experts demonstrably free from the possibility of influence. Those who criticize the present situation would do better to attack cuts in funding which force workers to seek alternative finance.

The second strand of motivation stems from the size and anonymity of modern industrial society. Large, seemingly remote corporations now produce nearly three-quarters of our food. They offer us admirable consistency and choice, no doubt, but we can easily feel that we are unimportant cogs in a vast machine. Consuming a different diet from one's fellow or, better, campaigning for changes in the national diet, is a rebellion against conformity, a statement of individuality. Although 90% of those claiming to be allergic to foods can be shown not to have a genuine allergy,[6] many angrily refuse to accept the diagnosis. Their 'specialness' has been taken from them and to abandon their campaigning zealotry is to merge back into the faceless mass. The same motivation, tinged with nostalgia, probably lies at the root of the desire for 'natural' products ('old fashioned country-style beefburgers made from 100% natural beef'). At the time when nearly everyone ate bread made from coarse, stone-ground flour, the trend-setters of the upper ten thousand ate the new white bread. There are sound reasons for consuming adequate fibre, but fashion is not one of them.

In summary, misinformation can arise at any point in the information chain and may stem from misunderstanding of the data or the desire of individuals to achieve some sort of prominence, the area of their endeavours being pure chance. Since belief is founded upon basic human needs, there is little point in producing logical counter-arguments after the misinformation has become a matter of faith. The best defence — as against so much in the way of demagoguery — is a public sufficiently educated and sophisticated to recognize nonsense when they see it. If this is so, a heavy responsibility is placed on education, both formal and 'popular'. The former is the subject of another theme of this conference; the latter is discussed in the remainder of this paper.

NUTRITION MESSAGES: WHAT MESSAGES? TO WHOM? FROM WHOM?

What Messages?

The most influential bodies in determining what nutrition messages should be promulgated have been NACNE[7] and COMA.[1] Health authorities that have instituted a nutrition policy have translated dietary goals into four key recommendations: (1) eat less fat; (2) eat more fibre; (3) eat less sugar and (4) eat less salt. There has been a tendency, however, in some of these programmes, to regard individual foods as 'good' or 'bad' when it is the overall contribution of the different foods to the diet that is important.[8] What is consumed on any particular day is often irrelevant; it is the long term average dietary patterns that matter.[9] Recommendations must describe good and bad diets with allowance for occasional consumption of (for example) high-fat products or they will become prescriptive.[8]

The state of nutrition science and the standing and public perception of nutritionists are likely to affect people's attitudes to nutrition and their receptivity to nutrition messages. It is, therefore, worthwhile briefly to examine this aspect, since one cannot escape the impression that the public perceive that nutritionists are scientists who never agree about what constitutes a proper diet and are frequently changing their minds. Nutrition is a new science. The discovery and definition of the nature of many of the essential components of the diet was at its height during the lifetime of many living nutritionists. Even so, we have seen many shifts in emphasis on what constitutes good nutrition. Boyd-Orr concluded that a significant proportion of the UK population was malnourished and advocated that to achieve optimal nutrition many people needed more 'protective foods', especially dairy products and fresh vegetables. Later, weight-watching was to become a widespread preoccupation and, for a period, concepts of energy control were dominated by the ideas of Yudkin[10] who postulated a 'carbohydrate craving' while fats were held to be self-limiting in terms of energy intake. Through the medium of magazines, newspapers and organizations dedicated to weight control, the public were imbued with the idea that foods in which refined sugar or starch predominated were taboo.

Then came NACNE,[7] with its emphasis on reduction in the proportion of energy taken as fat and refined sugar, and in the amount of salt consumed and its encouragement to increase correspondingly the consumption of bread and pasta to raise intakes of complex carbohydrates. NACNE was notable, not for its qualitative conclusions but for trying to quantitate its recommendations reasonably concisely. It also revealed quite interesting changes in attitudes to accepted concepts in nutrition. The term 'balanced diet' was regarded as outdated since it harked back to the days when we were more preoccupied with nutrient deficiencies. Yet it can be argued that 'balance' is equally applicable to problems of correcting nutrient excesses.[11] There is, one suspects, a hint of inconsistency here, in that many who argue against the term 'balance' on these grounds are the same who argue that modern food processing is a prime source of nutrient deficiencies.[12]

In the eyes of consumers, many of whom have seen a succession of such apparent changes in attitudes through their lifetimes, this is seen as nutritionists' vacillation. At least one unfortunate result is that, although well aware of current nutrition messages, people become disillusioned and uninterested in applying the principles to their own lives. In fact, uncertainties about the origins of the universe, the structure of subatomic particles or the nature of animal behaviour are no less prevalent but they excite little interest because they are not seen to impinge upon our daily lives. Scientists work by constructing hypotheses, testing them experimentally and reconstructing the hypotheses when tested to destruction and this is what we witness in the changing views of nutrition. The gulf lies in the failure to comprehend the scientific method, which says much about the state of scientific education in our schools. One conclusion is that initiatives in nutrition education can only be successful in the context of improvements in our educational system as a whole.

To Whom?

Recommendations for dietary change in the COMA report[1] were for individuals, not population groups. Of course, there are those who are more in need of dietary change than others and they can be identified by mobidity and mortality patterns or by poor eating habits, which in turn may be influenced by a number of factors as follows:

Different income and social class groups

Higher risk of mortality from coronary heart disease (CHD) occurs in lower income groups;[13] indeed, the poorer sections of society suffer more generally from ill health.[14] Despite the often presumed relationship between high fat intake and the incidence of CHD, lower income households tend to eat diets lower in fat than higher income households,[15] the bulk of foods eaten by the former group being cheap, starchy and filling, such as bread and potatoes.[16] In households with three children, fruit consumption is one-third less for those with low incomes compared with those with high incomes.

There is growing evidence that people with low incomes may not be able to afford to adapt their diets in line with nutritional guidelines, with the unemployed having the worst dietary standards.[17] Many 'healthier' foods are highly priced and are less available in low income areas. Nutrition educators face great difficulties in areas of multiple deprivation.[18] These areas are characterized by poor housing, high unemployment and general dereliction. For low income groups, nutrition education and information is largely ineffective.

Geographical differences

Within the UK there are regional differences in health patterns, especially in CHD incidence, the greatest incidence being found in Scotland and Northern Ireland. In England, regions with the greatest current CHD incidence seem to coincide with areas where infant mortality was greatest in the early part of this century.[19]

Age and sex differences

Females tend to consume a higher proportion of their total dietary energy as fat than do males, as do the elderly in general.[20] Younger people generally have more flexible dietary patterns and any modifications are likely to produce significant future reductions in diet-related diseases.[8] Nevertheless older groups could also benefit from dietary advice but are more resistant to change. It is a generally held belief that the elderly are set in their ways and will not try new food. Holdsworth and Davies[21] have described an effective plan for the nutrition education of the elderly and Bilderbeck and her colleagues[22] found that the elderly can be influenced by the persuasive powers of relatives, mass media and hearsay. Nevertheless, it may be necessary to change nutritional advice according to the target age group and there may be an age above which it becomes inappropriate.[8]

Groups with particular risk factors for CHD

People with diabetes, a family history of heart disease and those with several risk factors present concurrently (including obesity) are at higher risk than average of coronary heart disease. COMA[1] suggested that dietary change for these groups should be greater than for the general population.

A point that is relevant to this conference, concerned as it is with the whole food chain, is that surveys conducted to assess attitudes to, and understanding of, nutrition are without exception concerned with consumers. To our knowledge there is no information on attitudes and knowledge of producers, manufacturers, educationalists, administrators or legislators, all of whom have a powerful influence on the food chain. It is difficult to see how policy can be influenced if we are ignorant of the knowledge, attitudes and motivation of these other participants. A group that currently wields enormous influence is that of the retailers.

Insights into their motivation can be gleaned from accounts of the establishment of nutritional labelling and education programmes by supermarket chains.[23]

From Whom?

Government, local authorities and community programmes

Although the COMA recommendations to government were that nutrition education should be promoted, there has been little action on the part of government so far. In an attempt to translate the COMA recommendations for the general public, in 1985 the COMA guidelines were published in everyday language in *Eating for a Healthier Heart* by the Joint Advisory Committee on Nutrition Education (JACNE).[24] The booklet was not widely available, however, and was described as confusing by some.[25, 26] Until recently, most government funds available for nutrition education were channelled through the Health Education Council. The sums involved, however, were very small in comparison to the size of the task. (In 1986 the annual budget of the HEC was £30·5 m of which only £9·5 m was spent on dietary issues.)

In the meantime, although there has been no government directive obliging health authorities to take action on the recommendations in the COMA report, initiatives have been taken by some health authorities, dietitians and catering managers within the National Health Service. Indeed, in 1981, prior to the publication of the COMA report,[1] the DHSS Handbook *Care in Action* advised health authorities to draw up positive health promotion strategies. Apart from action by some forward-looking Borough Councils (e.g. Islington Health Authority) and nutrition policies operated for specific community groups such as infants, the elderly and ethnic minorities, by August 1985 only five of the fourteen regions in England and four of the fourteen in Scotland had, or were developing, a food policy.[2] In Wales, the Welsh Heart Programme has support from a diversity of organizations ranging from health authorities to industry and commerce. It concentrates mainly on personal education and changes in institutional catering, although food manufacturers and producers are being encouraged to modify their processes and products in line with the recommendations of the COMA report. In England, the 'Look after your heart' campaign does not seem to have gained the hoped for momentum.

The high cost and limited success of medicine successfully to treat cardiovascular disease provides impetus for preventative approaches to the problem. With the public already bombarded with misconceptions about nutrition from several sides (see pages 17–20) there is urgent need for a concerted effort in nutrition education. Government initiative is required and, perhaps, coordination as well.

The media

Since the guidelines were published, the media have allocated more time to food-related issues than ever before. The impact of television is immense: even a single programme can have a sustained effect on purchasing behaviour (e.g. skimmed and semi-skimmed milk.[2]) Similarly, books like *The F-Plan Diet*[27] have had great effect. This was a bestseller and undoubtedly contributed to growth in sales of products with a high fibre content.

Nevertheless, the mass media are not an extension of the education system, but part of an entertainment business whose primary aim may be to make money.[28] Unfortunately, media coverage of food and nutrition does not usually present a considered evaluation of all the available material. The time element may also preclude full evaluation of all the issues. The resulting assimilation of misinformation by the public has already been discussed.

It could be argued that information about the first stage of the food chain — the agricultural production of food — is provided in a more professional manner than information about food processing. Issues such as growth promoters and pesticide residues in agricultural products are no less emotive than food additives and cholesterol, yet it seems that programmes like 'Farming Today' manage to present agriculture in a professional way and, it is claimed, have an enormous following by the non-farming public. Should not food and nutrition programmes have the same degree of technical support?

Nutrition and medical professions

The provision of nutrition education for the public is not progressed by the lack of communication between nutritionists involved in research and education. There is a need to define the professional function of nutritionists by the creation of a professional body, within which their relationship with associated disciplines may be established. The proposed formation of special interest groups within the Nutrition Society, one of which could be concerned with education, may act as a spur to involving the nutrition profession more with public education. Although COMA[1] recommended that medical practitioners should be 'vigorous in identifying and advising people who have increased risk of coronary heart disease' so that they could be given 'special advice regarding diet', there is little evidence so far that this is happening and nutrition education within medical training remains rudimentary. In this connection, the announcement of two new Rank Chairs of human nutrition within British Medical Schools is much to be welcomed.

Food manufacturers and multiple retailers

Food manufacturers' development and promotion of new products and voluntary nutritional labelling has been partly responsible for the growth in awareness of diet and health issues. In changing the UK diet, this sector of the food chain faces greater difficulties than those faced by the retailers, as developing new products means a major upheaval, involving capital investment, retraining staff and obtaining suppliers. The shift in the balance of power in favour of the major retailers and accelerated rates of change in the food markets have increased manufacturers' difficulties.[2] Innovation is now an increasingly essential factor in the viability of food companies; larger companies are able to commit substantial resources to research and development while small and medium sized firms find it more difficult.

It could be argued that food manufacturing is a business and nutrition education is not part of the job. Some food companies, however, have entered the field of nutrition education by, for example, providing illustrative booklets. For example, Van den Berghs have set up the Flora Project for heart disease prevention and this is not unrelated to changes in their sales of Flora margarine. The National Dairy Council, on behalf of dairy producers and the dairy trade, has developed teaching materials in food and nutrition for use in schools. Richardson[29] advocates that 'any nutrition or ingredient information on a label should be backed up with an educational effort to ensure that consumers could apply their nutrition knowledge to make informed choices about food'. There is a danger in introducing a nutrition labelling programme with no educational programme to support it.

Perhaps the most significant development in recent years has been the emergence of the retail multiples as a major force in the food sector.[2] In the UK, a small number of supermarket chains now dominates food retailing. They can afford to conduct detailed market research to detect changes in consumer preferences. With this information, they can formulate precise specifications for their own private label products as well as for fresh fruit, vegetables and meat. Unfortunately, this information will not be generally accessible to nutrition policy makers, as it will remain confidential.

Before the supermarkets expanded, the retail sector consisted mainly of a large number of small shops which simply had to select their stock from what was available. Under these circumstances, agriculture and food manufacturing companies effectively controlled what foods were made available to consumers, with retailers having little direct influence.[2] Now, the supermarkets are extremely keen to respond to consumer preferences. This means that the modern consumer has more influence on food markets than ever before. Some food retailers (first Tesco and then Sainsbury and Safeway) have anticipated the 'healthy' diet boom and developed policies designed to help the consumer choose such diets. These have included nutrition labelling, stocking high-fibre, low-fat, low-salt and low-sugar foods, providing nutrition information and removing certain additives from own-brand foods.

How Should the Messages be Transmitted?

The first question is whether dietary guidelines are attainable? There is little point in advocating dietary change for the general public if they are not. Studies of the ability of dietitians (a well motivated and knowledgeable group) to eat a 'healthy' diet have shown varied success[30, 31] and have indicated that, especially for fat, difficulties are encountered, even by well informed groups. For the public, nutrition information should be clear, concise and relate to available foods and traditional diets. Success in delivering nutrition messages will depend on (a) frequent repetition, (b) aiming them at behavioural change and (c) involving the individual.

Changes in consumer attitudes, awareness and knowledge of nutrition

To predict the success of future programmes for nutrition education, evidence of changes in consumer attitudes to diet and health in response to earlier initiatives is required. Most studies have been concerned with consumer attitudes to diet, which is rather different from knowledge of nutrition. Assessing attitudes is more straightforward than assessing knowledge, since the former can be expressed in descriptive text, while the latter requires a form of quantitation. It is not surprising that most surveys record their findings in broad descriptive terms. In 1976 Sims[32] proposed that nutrition education programmes should adopt as a primary goal that of improving attitudes about nutrition as well as imparting nutrition facts and concepts.

In the three years since NACNE/COMA, few surveys have been reported, and only with a limited number of people[17, 33-44] (Table 1). There are few surveys before 1983/4 with which to compare these data. There was evidence of increasing knowledge and awareness between the early 1960s[45-47] and the early 1970s,[48] although the extent of knowledge, whether right or wrong, did not appear strongly to determine the foods consumed. Although there is clearly a need for further survey information as the guidelines are introduced, recent data indicate that consumer awareness of diet and health is still rising. In some instances, nutritional considerations rival price in influencing food purchase.[33] The public are becoming more familiar with nutritional terms and many are reasonably competent at associating them with specific foods.[2] Nevertheless knowledge tends to be superficial[33, 35, 48] and misconceptions abound. For example: all fat is thought to be visible; it is good to get a lot of protein; carbohydrates are 'starchy' and bad for you. Technical terms used by nutritionists (and rarely considered by them to be misleading) are confusing because they conflict with everyday usage. For instance, people think of acids as burning and damaging, so they do not know what to make of terms like 'fatty acid' and 'folic acid'. Some people think that saturated fat means saturated with fat, so that it might be possible to squeeze it out or shake it off. Energy is thought of as vitality and therefore highly desirable. It is not necessarily thought of in the same context as 'calories'. 'Calories make you fat and energy is a mark of vigour and health' — opposites! The concept of percentage of energy is obdurately incomprehensible.

Commercial promotion can create a high degree of awareness without necessarily increasing knowledge. Of those questioned by the Bradford University Food Policy Research Unit,[33] 87% agreed that 'polyunsaturated fat is better for you than saturated fat'. The link between the polyunsaturated nature of 'Flora' margarine and its sunflower seed oil content, stressed in advertising, has led many to interpret the less saturated fats advice as 'less animal fats'. Many health educators have been guilty of this oversimplification, as illustrated by a passage from the Health Education Council (HEC) booklet *Fat: who needs it?*:[49]

In particular, cut down on the solid, mainly animal (including dairy) fats — saturated fats as they are called.

Thus consumers receive messages that are either incorrect, as in the last example, or are strictly correct but leave an impression in the mind

TABLE 1
Recent Surveys of Consumer Attitudes to Diet and Health

Reference	Title	Coverage
Lang *et al.* (1984)[17]	*Jam Tomorrow?*	1. Questionnaire 2. Some 24 h recall surveys 3. Sample size: 1 000 low income consumers from N. England 4. Food consumption patterns 5. Attitudes to food 6. Food choice factors 7. Food consumption changes
Fallows and Gosden (1985)[33]	*Does the Consumer Really Care?*	1. Qualitative discussion groups 2. Quantitative questionnaires 3. Sample size: 132 respondents 4. Demographic and behavioural analysis 5. Attitudinal analysis 6. Ranking of foods by 'healthiness' 7. Further information and analysis available
Taylor Nelson Group (1985)[34]	*Survey of Attitudes to Food*	1. Interviews 2. Ten-day food diaries 3. Sample: 430 respondents from AB or DE socioeconomic group 4. Average height and weight 5. Knowledge about food 6. Attitudes to food 7. Food consumption data
MAFF, Consumers' Association, and National Consumers' Council (1985)[35]	*Consumer Attitudes to and Understanding of Nutrition Labelling*	1. Qualitative survey 2. Quantitative questionnaire 3. Sample: 820 interviews 4. Food consumption changes 5. Knowledge of food 6. Attitudes to food labels 7. Practical use of labels 8. Full data available
Freckleton (1986)[36]	*The Impact of a Supermarket Nutrition Information Programme*	1. Eight discussion groups with 68 participants 2. Unstructured with free discussion. Encouraged about food, nutrition, labelling and education. Qualitative

TABLE 1—*contd.*

Reference	Title	Coverage
Wright and Slattery (1986)[37]	*Talking about Healthy Eating*	1. Discussion groups of regular shoppers at a northern supermarket 2. No preplanned structure with conversation encouraged. Qualitative
Slattery and Wright (1986)[38]	*The Consumer Reaction to a Healthy Eating Initiative*	1. Group discussions before and four months after the introduction of the campaign 2. Open structured discussion about diet and health issues and attitudes to different foods. Qualitative
Phillips (1986)[39]	*Diet and Nutrition Information: A Survey of Attitudes and Knowledge in Reading*	1. Sample: 100 respondents 2. Quantitative questionnaire 3. Food habits 4. Food knowledge 5. Nutrition education and labelling
Wright (1986)[40]	*Consumer Attitudes to Healthy Eating: Part I*	1. Sample: 1 458 females 2. Quantitative questionnaire
Wright (1986)[40]	*Consumer Attitudes to Healthy Eating: Part II*	1. Sample: 60 supermarket shoppers
Health Promotion Research Trust (HPRT) (1987)[41]	*Health and Lifestyle Survey*	1. Quantitative survey 2. Sample: 9 003 3. Demographic and social circumstances 4. Physiological measurements 5. Cognitive function tests 6. Personality and psychiatric status 7. Risk factor analysis 8. Attitudes and beliefs about disease and health 9. Database available

(continued)

TABLE 1—*contd.*

Reference	Title	Coverage
D'Arcy, Masius, Benton and Bowles (1986, 1987)[42]	*The Healthy Eating study I 1986 II 1987*	1. Sample: 6 000. Limited information 2. Data from 192 attitudinal scales questionnaire 3. Cluster analysis of results produces segmentation of consumers by behaviour and attitudes to diet and health
Sheiham *et al.* (1987)[43]	*Food Values: Health and Diet*	1. Quantitative questionnaire 2. Sample: 3 100 3. General attitudes to diet and health 4. Demographic characteristics of healthy eating 5. Relationship between attitudes and diet 6. Changes in eating habits 7. Images of different foods
MAFF (1987)[44]	*Survey of Consumer Attitudes to Food Additives*	1. Qualitative survey 2. Quantitative questionnaire 3. Sample: 2 000 4. Concern about health 5. Changes in eating patterns and purchasing behaviour 6. Large database; good data disaggregation

which is likely to be false because the consumer has insufficient background knowledge.

Changes in eating patterns

On the whole, those who are least well informed are also those who show less dietary change, but the mere acquisition of knowledge does not mean that there will be behavioural change. Nutritional factors are not the only reason for food choice: image of caring wife and mother, time, money and cultural background all play a part. Only under special circumstances does nutrition become a high priority in food choice (e.g. pregnancy[50]). Children pose a considerable problem, as some women are reluctant forcefully to introduce alternative foods for their families.[40]

Monitoring the impact of specific health initiatives is not easy, but attempts have been made to monitor sales of 'healthy' foods in supermarkets (discussed by Wheelock *et al.*[2]) and to hold group discussion with shoppers[40] with some success. Studies such as these show that in the UK, as in the rest of Europe, the desire for quality and an increasing awareness of health issues are becoming important considerations in food selection. This is seen in anomalies in purchasing behaviour where the demand for certain foods has developed independently of changes in retail prices. Examples include consumption of liquid milk falling despite a fall in the real price, and consumption of filleted white fish increasing with a rise in real price.[2] Products categorized as 'low fat' are selling well. The effects of changes in food choice can be followed for fat consumption, which was the main issue highlighted by COMA.[1] Total fat consumption has declined over the last ten years but its contribution to total dietary energy has risen, with large falls in carbohydrate, causing significant falls in energy intake. Saturated fatty acid intake has fallen from a 1959 peak of 57 g/person/day to 44 g in 1982.[1] Polyunsaturated fatty acid intake, however, has increased from 9 g/person/day to 12 g in 1982, so changing the polyunsaturated/saturated (P/S) ratio from 0·17 to 0·27 over this period.[20]

To what extent these changes are a result of, or influenced by, nutrition education is uncertain. Writing in 1965, McKenzie and Mumford[51] wrote:

It would be wrong to regard nutrition education as always being an effective tool for modifying food habits. It would be equally wrong, however, to regard it as always being ineffective. The work done so far seems to suggest that the success or failure of an education programme depends upon the methods used, the personalities involved and the circumstances prevailing in the area at the time ... There is a particular need for social scientists to work in close cooperation with nutritionists in order to establish a satisfactory and simple methodology and to encourage the use of evaluation as an integral part of any programme they design.

Certainly we can expect to make little progress with nutrition education programmes unless we have effective means of evaluating them and of targeting them according to the knowledge, background and circumstances of the recipients.

The next sections are concerned with the role of the food manufacturer and retailer in the whole exercise of informing consumers about the nutrient content and the nutritional attributes of foods and the extent to which current methods are effective or could be improved.

FOOD PROCESSING: RESPONSES TO THE GROWING HEALTH AND FITNESS SEGMENT

Recent developments in the market-place have witnessed a remarkable growth in products stating perceived nutritional benefits such as less fat, less sugar, fewer calories, less salt, high content of essential poly-unsaturated fat, enriched with vitamins, minerals, fibre, calcium, freedom from additives, etc.[52] Fortunately, promotions for the vast majority of products in the rapidly growing 'healthier eating' markets draw attention to overall quality and adhere to the more responsible approaches to health and nutrition. In some cases, however, claims have implied improvements in health that cannot be fully substantiated and there has been a huge increase in 'negative' and 'comparative' claims that can lead to technical and legal difficulties as well as confusion.[53]

This section discusses concerns and apparent inconsistencies in the efforts of the food industry to respond to customer demands for healthier eating. Generally, the food industry's policies and actions on food additives and nutrition reflect a genuine desire to scrutinize the composition and nutritive value of processed foods and to develop marketing and promotional strategies which will permit the consumer to make informed choices and enjoy the benefits of all that modern food technology can provide.

Promotional Measures

How foods are described, promoted or advertised is the result of commercial decisions made on commercial grounds, within the constraints of what is required or prohibited by the laws of the land. Sections 6.1 and 6.2 of the Food Act 1984 render it an offence to label food or to publish an advertisement for food which '(a) falsely describes the food or (b) is calculated to mislead as to its nature, substance or quality'. The use of any terminology must be seen to be understood by the ordinary person (on the Clapham omnibus!). The use of scientific or supposedly scientific, facts, issues, judgements or implications to describe, promote or advertise foods, requires very special care and attention. Much of present day food advertising in print and on television deals with aspects of food quality, especially food safety and nutrition, either directly or indirectly. Most of the copy addresses legitimate consumer concerns and describes how food products can fulfil a legitimate consumer need or desire.

Unfortunately, many sectors of the media and the food industry are exploiting ill-defined public fears and ignorance about nutrition and food additives and it is up to responsible manufacturers and retailers not to add fuel to the fire and lay the food industry open to further criticism. Immediate marketing gain, essential though it may be in a competitive environment, is only one factor in the long range economic health of the food manufacturing industry.

Negative Claims

It is a legitimate promotional measure to proclaim the merits of a particular food product or the method of its manufacture, provided that the information is accurate and the claims can be justified. Any claim, however, which denigrates food ingredients whether overtly or by innuendo, should be thoroughly scrutinized by the marketing, technical, legal and public relations functions of the individual companies. Over the last two years, some irresponsible promotional campaigns have stated and implied that foods containing certain permitted additives are, *per se*, less wholesome or less safe, or less healthy than those without the additives. Conversely, it has been stated and implied that foods without certain permitted additives are, *per se*, more wholesome and safer, or healthier than those with the additives.

Nevertheless, negative claims such as those outlined in Table 2 are often a useful aid to inform the consumer of the absence or non-

TABLE 2
Types of Negative Claims
(Adapted from Working Paper Alinorm 85/22A, Appendix IX)[55]

A. Claims which indicate complete absence:

      e.g. *no_____*   *contains no_____*
      free from_____   *_____free*
      non-_____   *un-_____   *_____-less
      without_____

B. Claims which indicate qualified absence:

      e.g. *no added_____*   *no_____added*

---

addition of substances in foods. Care must be taken to use these potentially useful statements in such a way that they are not false or misleading. The proposed restriction, by the Food Advisory Committee,[54] of the claim that a food is 'free from' one category of additive when another category or ingredient having broadly similar effect is used, ignores the fact that consumers are not uniformly concerned or knowledgeable about all classes of additives. Although it may be logical to forbid a manufacturer from claiming that a soft drink is 'free from tartrazine' when it is still coloured with another azo dye (e.g. Ponceau 4): there is no reason why a negative claim about a colour should not be made when the drink contains e.g. a preservative for microbiological stability. It has been said that negative claims may mislead when a certain additive or ingredient is not permitted or when the additive or ingredient is not normally present, either in the product or in a class of products. It is sometimes important, however, to draw the consumer's attention to the fact that a certain product does not contain a substance, particularly if this fact could change purchasing behaviour and perhaps stimulate a product sector. For example, canned goods generally do not contain preservatives; evaporated milk does not contain added sugar (sucrose).

Special care and attention must also be paid to expressions such as 'sugar free' and 'no sugar added' where products could contain sugars other than sucrose, e.g. the milk sugar, lactose, present in an ingredient such as skim milk powder. These, and similar expressions, should only be used if the information is followed immediately by an equally clear statement indicating which sugars are present or have been added to the food.

The subject of negative claims is being debated by the Codex Committee on Food Labelling with regard to the possible revision of the Codex General Guidelines on Claims.[55] The discussions take into account the type of claims shown in sections A and B of Table 2. The types of claims in section A are clear cut in that the substances to which they refer should not be present in the product. Statements and claims would necessarily have to take into account any possibility of the substances in question being introduced into the food directly through other ingredients by means of carry over (which is particularly important if the carry over components can cause intolerance). The claims expressed in Section B imply that the substances may be present in the product in its natural state although none has been added during preparation. Such claims would not be valid if the substance was added indirectly through other minor ingredients (e.g. salt in mixed spices) apart from any naturally occurring level.

Negative claims have also been utilized to highlight the absence of certain substances in foods which are intended for specific dietary regimens. These include statements such as 'gluten free', 'lactose free', 'non-alcoholic', etc. More recently, the Leatherhead Food Intolerance Database has extended this area of dietary concern to include specific food additives which can have particular adverse health connotations e.g. absence of the food colour, tartrazine. Religious beliefs, vegetarian and vegan lifestyles have also resulted in claims to aid the selection of foods to meet these dietary restrictions, e.g. 'contains no pork', 'no animal fat'.

There are many ways to control the use of negative claims; any new guidelines should enable the food industry to continue to develop claims which are useful and meaningful aids to consumers and restrict the use of those which are spurious and emphasize qualities that are only marginal and which may, therefore, give a completely wrong impression of the food and its use.

Naturalness

The meanings of the terms 'natural', 'artificial' and 'nature-identical' are open to many interpretations, and the government, through the Ministry of Agriculture, Fisheries and Food (MAFF) asked the Local Authorities Committee on Trading Standards (LACOTS) to investigate the subject in 1986. Their report[56] has now been published and was used in the preparation of the Food Advisory Committee (FAC) statement[54] on the use of the word 'natural' and similar terms in the labelling, advertising and presentation of food. MAFF is now in the process of collecting comments from interested parties. The Committee on Toxicology (CoT) has also indicated to the FAC in their *Report on the Review of the Colouring Matter in Food Regulations*[57] that there is no inherent reason why chemicals present in nature, for example, the so-called 'natural' and 'nature-identical' food colours should be safer than any others. All added colouring matter, whatever its origin, needs to be examined for both need and safety-in-use. This philosophy applies to all food additives.

There is little case law now governing the use of the word 'natural' or related terms. In the 'Britvic' case, it was argued by the prosecution that the description 'natural' applied to orange juice that had been concentrated, pasteurized and subsequently reconstituted with water and that the term 'natural orange juice' was false under Section 6.1.a of the Food Act. However, both the magistrates and, on appeal, the High Court, ruled that no offence had been committed. In giving their verdict, the magistrates stressed the absence of additives in the product.

The term 'natural', however, has been abused, and a practical code of practice will certainly help clarify the position. Although the debate is still intense following the publication of the FAC statement, there are areas of the report that may be impractical. Generally, the principles for foods to be advertised or labelled as 'natural' should be that they are derived from material that occurs in nature; the products, and/or the product ingredients should be in their original state, or be minimally processed by a process used to make the food edible or to make it safe for human consumption; the product should contain no 'nature-identical' or 'artificial' ingredients and contain no other natural additives not normally present, e.g. beetroot as a colour for raspberry yoghurt. The physical or food preparation techniques should be clearly defined in the FAC report to minimize confusion.

There is at present, no statutory definition within the UK, EEC or the Codex Alimentarius of the term 'nature-identical'. For example, substances of this type are identified and then chemically synthesized so that they are indistinguishable (both chemically and biologically) from their naturally occurring counterparts, e.g. β-carotene, canthaxanthin and riboflavin. Artificial ingredients are those substances made up from chemical compounds that have not to date been identified in nature.

Most scientists and legislators would agree that the use of the terms 'natural' and 'naturalness' should be carefully controlled and that each product formulation and process be assessed on its own merits, bearing in mind how the ordinary member of the public would view the use of the words. It should, however, be noted that some of the largest food manufacturers have received no customer complaints about the use of the word 'natural' and that the potential for misleading the public may be really quite small.

Claims Resulting from the Fortification of Foods with Vitamins and Minerals

The Food Labelling Regulations, 1984, do not include vitamins, minerals and other nutrients as 'additives' insofar as they are used solely for the purposes of fortifying or enriching food or of restoring the constituents of food.

Several nutrients have been added to food and drink products around the world as public health measures and as a cost effective way of ensuring the quality of the food supply. Addition of some nutrients has also formed the basis of marketing strategies in product development.

The main criteria for selecting nutrients to be added to food are that they are shown to be necessary, safe and effective. Nutrition enrichment of foods can help prevent nutritional inadequacies in a population where there is a risk of deficiency and when intervention is needed to correct a proven deficiency in an identified segment of the population.

Effectiveness of a fortified food is influenced by whether the food that is to carry the nutrient is going to be acceptable, consumed by those who need or want it and be at a price they can afford. The nutrients must be bioavailable and sufficiently stable under normal — and perhaps unusual — conditions of storage and household use.

Care must be taken to ensure that consumption of foods containing added nutrients will not create a nutritional imbalance and that an excessive intake of the nutrients will not occur, bearing in mind the cumulative amounts from other sources in the diet.

Food fortification requires careful attention to food regulations, labelling, nutritional rationale, cost, acceptability of the product to consumers and a careful assessment of technical and analytical limitations for compliance with label declarations. The indiscriminate additions of nutrients to foods, however, should be discouraged and information on food labels should not overemphasize or distort the role of a single food or component in enhancing good health.

Health Claims

Implicit claims for the presence of vitamins, minerals, proteins, etc. are already controlled by the Food Act 1984. Claims as to the suitability of a food for use in the prevention, alleviation, treatment or cure of a disease, disorder or particular physiological condition are prohibited unless they follow strict rules for such foods as those for special dietary uses. More recently, however, there has been a trend towards more explicit health-related and disease prevention claims on food pack — e.g. calcium and osteoporosis, and dietary fibre and cancer prevention.[58]

While health claims on food labels can be an important way of conveying nutrition information to the public, there is concern that, without sufficient control or guidance, such claims would potentially furnish misleading and/or harmful information. There is a need to develop a case by case basis to evaluate any health claim. The basic problems are how to allow valid, appropriate health claims on foods without opening the door to misleading and fraudulent claims and to ensure that disease prevention claims are founded on and are consistent with, widely accepted, well substantiated, peer-reviewed scientific publications. Any health statements on food labels should be consistent with those held by qualified experts. Further complexities in implementing health claims on food labels are related to the amount and kind of scientific data necessary to substantiate such claims (e.g. for a food containing adverse as well as beneficial components), the difficulty in simplifying complex health messages to fit the limited space on labels and the threat of 'power races' among food companies to gain a competitive edge. To maximize their effectiveness, health claims need to be used in conjunction with other nutrition education efforts.

Comparative Claims

MAFF is aware of the difficulties and potential confusion arising from the increasing use of 'comparative claims' and are already assessing ways and means of controlling them. Except for claims such as 'low' or 'reduced energy', which are controlled by the Food Act 1984, most other statements on 'high', 'low' or 'reduced' are either the subject of guidelines (e.g. quantitative fat levels to describe low-fat and very low-fat yoghurts) or manufacturers and retailers have based their comparisons for 'low' and 'reduced' on the so-called regular product and the healthy product containing 50% less and 25% less respectively of the component in question, e.g. low sodium. Confusion would arise from a 'low' sodium claim on one product, e.g. a dehydrated soup containing 10–15% by weight common salt, which is higher in sodium, compared with a low sodium claim on, say, a packet of crisps. Moreover, as more and more products claim 'low' or 'reduced', the goal posts move, and it is necessary to review and resubstantiate claims on pack. For example, five years ago standard pork sausages contained 30–32% fat; today they contain 22–25%; hence 'half the fat' has moved from 15% to 12%.

A number of international and independent committees have attempted to develop a rational approach to comparative claims, but so far, none has fulfilled the basic criteria for effective labelling. These criteria include a system of labelling that is: consumer orientated and meaningful, helpful at the point of purchase, allows people to select balanced diets and enables individual consumers to follow recommended diets. Some good efforts have been made to 'band' foods according to the quantitative amounts of fat, sugar, salt and fibre present per 10 MJ energy (see Table 3), per 100 ml or per 100 g as sold, etc.[59] In the United States, certain descriptive terms of sodium labelling have already been

TABLE 3

Summary of Recommendations on Banding Proposed by the Nutrition Advisory Committee of the Coronary Prevention Group (January 1987)[59]

Nutrient	High	Medium–high	Medium–low	Low
Total fat % energy	53 and over	35–52	17·5–34	17·5
Saturated + *trans* fatty acids % energy	22·5 and over	15–22·4	7·5–14	7·5
Total sugars: g/10 MJ energy	112 and over	75–111	37–74	37
Salt: g/10 MJ energy	7·5 and over	5–7·4	2·5–4·9	2·5
Fibre: Old method — g/10 MJ energy	45 and over	30–44	15–29	15
New method — as non-starch polysaccharide	27 and over	18–26	9–17	9

introduced into food law.[60] They are based on defined quantities of sodium in a serving of food. This approach is much more rational and provides a more scientific basis for descriptive claims (see Table 4). The use of absolute values is also consistent with the existing definitions for low-energy foods in the Food Act 1984.

A 'banding' system, providing clear nutrition messages, would be

TABLE 4

Sodium Descriptors in the USA

(Food & Drugs Administration; 21 CFR Chapter 1, 4.1.86)[60]

Sodium free	Foods containing less than 5 milligrams of sodium per serving
Very low sodium	Foods containing 35 milligrams or less of sodium per serving
Low sodium	Foods containing 140 milligrams or less of sodium per serving
Reduced sodium	Foods formulated to serve as and represented as direct replacements for foods containing at least four times the sodium content (75% reduction). The label has to bear information comparing the sodium content per serving with that of the food it replaces

helpful to consumers so long as it is not oversimplified and is based on scientific and quantitative principles. The consumer would be further confused by the use of a number of unspecified methods to describe foods. To date, if a comparison has been made, e.g. lower fat, a statement explaining the comparison has been included on the label declaration. It is, however, desirable from the food industry's point of view to have clear guidelines to ensure fair trade and avoid criticism of self interest in the choice of banding and the selective way it can be used on labels.

The prime nutrition education objective is not to identify so-called 'good' or 'bad' foods but good or bad diets. 'Banding' and 'traffic light' systems, unless properly developed, could undermine the longer term studies that are needed to find an effective means of educating consumers about food and nutrition.

In summary, food labelling and, in particular, the use of nutrition and health claims, is potentially the most significant food policy issue. Food labels will undoubtedly become one of the most widely read sources of information, and the gradual emergence of information, which goes beyond the factual quantitative data about nutrition, stems from the growing body of scientific evidence linking diet and health.

Food manufacturers are playing an increasingly important role in the trend towards 'healthier' eating by offering the consumer alternative products which taste as good as the original. New 'healthier' products tend to reinforce the concern for sensible eating on the part of the consumer. The current changes in the factors influencing consumer choice should provide new and interesting product developments and marketing opportunities.

NUTRITION LABELLING: A SYSTEM DESIGNED TO INFORM, EDUCATE AND MOTIVATE

If nutrition labelling is intended to be an instrument which consumers can use to implement dietary advice should they wish to do so, it follows that consumers should be able to understand it. To do this they may need to be educated about how to use it and to be informed about its purpose.

Certainly the majority of surveys and studies conducted to assess consumer understanding and use of nutrition labelling, suggest that better educated consumers are more likely to be able to understand and, therefore, use the information supplied.[61-66] Several of these and a number of other surveys have also indicated, however, that although education leads to greater awareness and knowledge of nutrition, diet and health and of the uses of nutrition labelling, this does not necessarily result in a behavioural change. It does not automatically motivate consumers to use nutrition information to alter their consumption habits and therefore their diets.[2, 36, 67–71] What is needed therefore, is a system that consumers not only understand but which they will actually be motivated to use. One of the commonest and most successful ways of doing this is to simplify the messages involved, thereby encouraging more people to participate.

One approach to the problem is to use an extended series of claims or similar ranking techniques. Evaluation of this type of system has shown that consumers can assess a product rapidly and make an appropriate decision to buy it.[2, 35a, 36, 71] This approach also has its disadvantages, however. Assessment of products individually can often lead consumers to perceive the product as 'good' or 'bad', 'healthy' or 'unhealthy', rather than to evaluate its contribution to the total diet. The introduction of such a system might therefore perpetuate this concept. In addition, this system makes no provision for the more nutritionally sophisticated consumer who may require more detailed information so as to devise diets rather than using nutrition labels simply to compare one product with another as is the case for the less sophisticated consumer.

To be able to assess their diets and make appropriate alterations, consumers need numerical information. For some consumers, who perceive numerical data as complex and time consuming, this is likely to be demotivating. There is thus a case for developing a dual system: one element permitting a rapid assessment of the product for comparative purposes, the other providing more detailed information for diet planning.[72]

Numerical information

The first consideration is the scope of the nutrition information to be included. How many nutrients? In keeping with the philosophy, expressed elsewhere in this report, that consumers will only be motivated to use the labels if the information is kept as simple as possible, we are advocating that numerical information should be limited to protein, carbohydrate, fat and energy. This information may then be expressed in a number of ways (see Table 5):

(i)   as percentage of energy contribution
(ii) as percentage by weight

TABLE 5

Nutrition Labelling Illustrating the Declaration of Nutrition Information Expressed Numerically.
(Calculated for women, aged 18–54 years, most occupations.)[72]

Nutrition information	As a percentage of the energy provided (%)	As a percentage of the weight of the product provided (%)	Nutrient density	Weight per 100 g of the product (g)
Fat	44	24	0·14	24·1
Protein	6	7	0·55	6·8
Carbohydrate	50	64	0·25	66·5
Energy	2 071 kJ/493 kcal	2 071 kJ/493 kcal	2 071 kJ/493 kcal	2 071 kJ/493 kcal

(iii)   as nutrient density
(iv)   as weight per specified weight (or volume/volume)

Of these, weight per specified weight has proved most popular and best understood by consumers, the others being viewed as complex.[35a,b] Regarding the basis to be used for this specified weight, consumer preference is less clear and appears to be dependent on the product.[2, 36, 71] For example, if the declaration refers to a discrete item, (to be eaten by one person on one occasion) it would be most relevant to provide information per serving. For all other declarations, where a 'serving' is not readily assessed, presentation of the information per 100 g would be most practicable. It is important to standardize the approach adopted and to recognize the difficulty of defining servings. Thus we suggest that in almost all cases the data be required to be presented per 100 g, but to permit additional declarations per serving where appropriate. Where the product is a small package, the contents of which are intended to be consumed by one person at one sitting, the 'per serving' declaration only would be sufficient. The alternative of presenting data in terms of energy units may be attractive to those familiar with the concept of calorie counting but has not been formally tested (Table 6).

Non-numerical information

This system is designed to allow rapid assessment and comparison of one product with another. The essence is simplicity, to motivate those consumers who are bewildered by more complex numerical systems and would be likely to lose interest. The options are (see Fig. 2):

(i)   verbal
(ii)   use of symbols to indicate contribution of nutrients by the product
(iii)   graphical, either bar chart or pie chart

TABLE 6
Nutrition Labelling Illustrating Nutrition Information Expressed Numerically as Units of Energy[72]

Nutrition information	kJ per 100 g	kcal per 100 g
Fat	912	217
Protein	113	27
Carbohydrate	1 046	249
Total energy	2 071	493

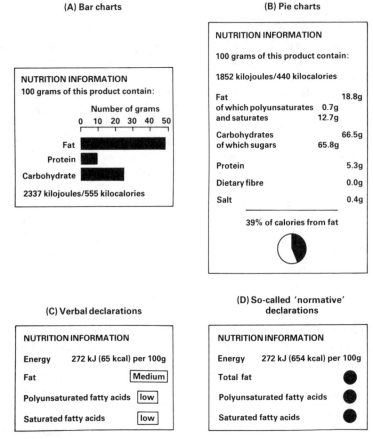

Fig. 2. Nutrition labelling illustrating various non-numerical systems.

Of these, graphical declarations have proved to be most popular with and best understood by consumers, the others being considered too judgemental[35a] or confusing once studied.[35b] Moreover, pie and bar charts can be used not only to illustrate basic data but to present more detailed information as depicted in Fig. 2. Again, the basis for presentation could be as percentage of energy; percentage of weight; weight per specified serving or as units of energy. The latter is suggested, since it not only allows consumers rapidly to assess the energy contribution of the macronutrients relative to each other but would also perform an educational function, illustrating the larger quantity of

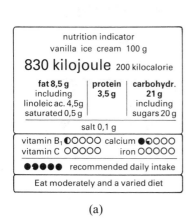

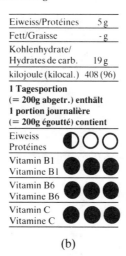

(a)   (b)

FIG. 3. Nutrition labelling illustrating the pictorial representation of micronutrients in (a) The Netherlands and (b) Switzerland.

energy contributed, weight for weight, by fat. Energy units are preferred as being better liked and understood by consumers than percentages.

The next problem is how to express the micronutrients should it be thought desirable and appropriate to declare them. In its second report on claims and misleading descriptions, the Food Standards Committee recommended that vitamins and minerals be expressed as a fraction of the RDA, this having more meaning to consumers than a simple declaration of weight.[73] Consumer testing, however, suggests that they are confused by the expression '% RDA'.[70] Attempts to avoid such confusion have included pictorial systems used by organizations in Switzerland and The Netherlands[72, 74] (Fig. 3). Such systems are cumbersome and tend to limit the number of nutrients that can be expressed, so that a numerical system is in the end likely to be most successful, though considerable thought needs to be given to the presentation method.

The dual systems illustrated in Fig. 4 should permit a variety of functions to be performed according to consumers' requirements, providing sufficient information for the more nutritionally sophisticated consumer and motivating the less sophisticated to become more involved with nutrition labelling. Its success is dependent on there being adequate education in its purpose and use. Motivation, however,

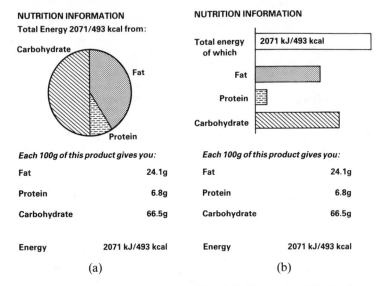

FIG. 4. Proposed dual system of nutrition labelling using (a) pie chart and (b) bar chart.

is the most important factor and there are many signs that this is increasing.

CONCLUSIONS AND RECOMMENDATIONS

In the UK, the NACNE[7] and COMA[1] reports have made a significant impact in changing attitudes towards diet in the UK. The acceptance by the government of the COMA report means a commitment on its part to encourage changes in the UK diet through one or more of the following routes:

1. Provision of nutrition education and information.
2. Regulatory policy: labelling regulations, compositional standards, advertising, etc.
3. Provisions strategy: regulating quality of food supplied for government institutions.
4. Research and development: development of new 'healthy' foods.
5. Pricing policy: taxation and subsidies.

Some of these routes, however, would lead to conflict with other policies, particularly with respect to current EC practice. Provision of nutrition information and education is less contentious and was specifically mentioned as a responsibility of government by COMA.[1] At the same time there is need for an education programme to support the future nutrition labelling (whether mandatory or optional); a development that was also encouraged by COMA.[1]

How can nutrition knowledge and its translation into healthy eating be improved? Until we have good baseline knowledge on what people already know, believe and, perhaps more importantly, what they want to know, we cannot begin to develop effective education programmes. (One schoolchild is quoted as saying: 'Don't teach us what you want to teach: teach us what we want to know!').[50]

The second problem is in agreeing what are the correct messages to be conveyed in any education programme. The basic tenets of what constitutes healthy eating are not too contentious among professional nutritionists. Nevertheless, all parties need to agree to avoid the unthoughtful use of the terms: 'good' and 'bad' foods; 'saturated' and 'unsaturated' fats (when it is only the *fatty acids* that can be so described); 'animal' and 'vegetable' fat in a misleading context. More thought needs to be given to the language of communication. Terms that are readily understood by professionals ('saturated', 'energy' and 'fatty acid' for example) are clearly going to perpetuate misconceptions in the lay mind and we need to take greater care in the presentation of information in lay terms.

We need to give greater attention to the targets of our educational programmes. The attitudes of consumers are strongly determined by their upbringing[50] and the more ingrained, the more difficult they are to change. Perhaps we should only be taking the long term view and concentrating our resources on the young who are tomorrow's food purchasers. Initiatives in nutrition education can only be successful in the context of the education system as a whole. More emphasis on the scientific method and on the practical value of science in the community is needed. Only then can the broadening of syllabuses to include nutrition and health as a universal component of the schools' curricula be effective.

In conclusion, sufficient resources from public funds should be set aside to embark on a major information and education campaign, aimed at school children, the medical profession, especially medical students, and the public as well as a means for monitoring the success of these programmes. So far, the government has not become greatly involved in nutrition education, but they may be forced into action as NHS costs increase. Nutritional issues are already having an impact on consumers' food choice and this trend is likely to accelerate as the nutrition messages are reinforced through product development and marketing. It is increasingly important that there is unity and cooperation between the different sectors of the food industry, with responsible marketing and advertising. A comprehensive approach to the food chain is required, with an integrated health policy involving agricultural and economic policy at national and international levels.

ACKNOWLEDGEMENTS

The authors are grateful to their many colleagues for helpful comment and discussion and especially to Mr S. Henson for preparing Table 1.

REFERENCES

1. COMA (Committee on Medical Aspects of Food Policy) (1984) *Report on Health and Social Subjects 28,* Department of Health and Social Security, HMSO, London.
2. Wheelock, J. V., Frank, J. D., Freckleton, A. M. and Hanson, L. (1987) *Forecasting and assessment in science and technology: Occasional papers, No. 130,* Directorate General for Science Research and Development, Commission of the European Communities, Brussels.
3. Wheelock, J. V. (1986) *Food Additives in Perspective,* Food Policy Research Unit briefing paper, Horton Publishing Ltd, Bradford.
4. Porter, H. (1984) *Lies, Damned Lies and Some Exclusions,* Chatto & Windus, London.
5. Roberts, H. R. (1978) *Federation Proceedings,* **37,** 2575.
6. Ferguson, A. and Barnetson, R. S. T. C. (1984) *Chemistry and Industry,* 6 Feb., 100.
7. NACNE (National Advisory Committee on Nutrition Education) (1983) *A discussion paper on proposals for nutritional guidelines for health education in Britain,* Health Education Council, London.
8. Gibney, M.J. (1985) In: *The Role of Fat in Human Nutrition,* F. B. Padley and J. Podmore (Eds), Ellis Horwood, Chichester, 173–81.
9. Marr, J. W. (1983) *British Nutrition Foundation Nutrition Bulletin,* **8,** 65–72.
10. Yudkin, J. (1975) *Chemist & Druggist,* 22 Feb., 259.
11. Conning, D. (1987) *BNF Comment,* 8 May.
12. Walker, C. and Cannon, G. (1984) *The Food Scandal,* Century Publishing, London.

13. Sanderson, M. E. and Winkler, J. T. (1983) *Lancet*, **2**, 1351-6.
14. Department of Health and Social Security (DHSS) (1980) *Inequalities in Health. Report of a Research Group chaired by Sir Douglas Black*, HMSO, London.
15. DHSS (Department of Health and Social Security) (1978) *Prevention and Health: Eating for Health*, HMSO, London.
16. Cole-Hamilton, I. and Lang, T. (1986) *A Report on the Impact of Poverty on Food*, London Food Commission, London.
17. Lang, T., Andrews, C., Bedale, C., Hannon, E. and Hulme, J. (1984) *Jam Tomorrow?* Food Policy Unit, Manchester Polytechnic.
18. Angrove, A. (1984) *Human Nutrition: Applied Nutrition*, **38A**, 5-16.
19. Barker, D. J. P. and Osmond, C. (1986) *Lancet*, **1**, 1077-81.
20. James, W. P. T. J. (1985) In: *The Role of Fat in Human Nutrition*, F. B. Padley and J. Podmore (Eds), Ellis Horwood, Chichester, 9-22.
21. Holdsworth, D. and Davies, L. (1982) *Human Nutrition: Applied Nutrition*, **36A**, 22-7.
22. Bilderbeck, N., Holdsworth, M. D., Purves, R. and Davies, L. (1981) *J. Human Nutrition*, **35**, 448-55.
23. Mason, T. (1986) In: *Food Policy Issues and the Food Industries*, J. Burns and A. Swinbank (Eds), Food Economics Study No. 3, University of Reading, 150-70.
24. JACNE (Joint Advisory Committee on Nutrition Education) (1986) *Eating for a Healthier Heart*, HMSO, London.
25. O'Donnell, M. (1985) Eschewing the Fat, *Guardian*, London, 30 Oct.
26. Bender, A. E. (1986) JACNE — under scrutiny, *Modus*, Jan., 33.
27. Eyton, A. (1982) *F-Plan Diet*, Penguin, Harmondsworth.
28. Turner, S. (1984) *Proc. Nutr. Soc.*, **43**, 217-18.
29. Richardson, D. P. (1985) In: *Food Policy Issues and the Food Industries*, J. Burns and A. Swinbank (Eds), Food Economics Study No. 3, University of Reading.
30. Black, A. E., Ravenscroft, C. and Sims, A. J. (1984) *Human Nutrition: Applied Nutrition*, **38A**, 165-79.
31. Cole-Hamilton, I., Gunner, K., Leverkus, C. and Starr, J. (1986) *Human Nutrition: Applied Nutrition*, **40A**, 365-89.
32. Sims, L. S. (1976) *J. Nutrition Education*, **8**, 122-5.
33. Fallows, S. and Gosden, H. (1985) *Does the Consumer Really Care?* Food Policy Research Unit report, Horton Publishing Ltd, Bradford.
34. Taylor Nelson Group (1985) *Survey of Attitudes to Food*, Taylor Nelson Group, Epsom, UK.
35. Ministry of Agriculture, Fisheries and Food, Consumers' Association and National Consumers' Council (1985) *Consumer Attitudes to and Understanding of Nutrition Labelling.*
    (a) *Summary Report — Qualitative Stage:* Susie Fisher Research Association, London.
    (b) *Summary Report — Quantitative Stage:* British Market Research Bureau Ltd, London.
36. Freckleton, A. M. (1986) *The Impact of a Supermarket Nutrition Information Programme*, Food Policy Research Unit report, Horton Publishing Ltd, Bradford.

37. Wright, G. and Slattery, J. (1986) *Talking about Healthy Eating*, Food Policy Research Unit report, Horton Publishing Ltd, Bradford.
38. Slattery, J. and Wright, G. (1986) *The Consumer Reaction to a Healthy Eating Initiative*, Food Policy Research Unit report, Horton Publishing Ltd, Bradford.
39. Phillips, C. (1986) *Diet and Nutritional Information: A Survey of Attitudes and Knowledge in Reading*, Food Economics Study No. 2, University of Reading.
40. Wright, G. (1986) *Consumer Attitudes to Healthy Eating:* Part I — text; Part II — statistical data, Food Policy Research Unit report, Horton Publishing Ltd, Bradford.
41. Health Promotion Research Trust (HPRT) (1987) *The Health and Lifestyle Survey*, HPRT, London.
42. D'Arcy, Masius, Benton and Bowles (1986, 1987) *The Bandwagon's Further Progress*, D'Arcy, Masius, Benton & Bowles, London.
43. Sheiham, A., Marmot, M., Rawson, D. and Ruck, N. (1987) In: *British Social Attitudes, Report of Social and Community Planning Research*, R. Jowell, S. Witherspoon and L. Brook (Eds), Gower Publishing, Aldershot.
44. Ministry of Agriculture, Fisheries and Food (MAFF) (1987) *Survey of Consumer Attitudes to Food Additives*, HMSO, London.
45. Brown, A. M., McKenzie, J. C. and Yudkin, J. (1963) *Nutrition*, **17**, 16–20.
46. Jenkins, N. K. (1964) *Nutrition*, **18**, 115–20.
47. Jenkins, N. K. (1964) *Nutrition*, **18**, 155–9.
48. British Nutrition Foundation (1973) *Food and Nutrition: Report on a Survey of Housewives' Knowledge and Attitudes*, BNF, London.
49. Health Education Council (1985) *Fat: Who needs it?* Health Education Council, London.
50. Thomas, J. (1980) In: *Nutrition and Lifestyle*, M. R. Turner (Ed.), British Nutrition Foundation, Applied Science Publishers, London, 157–67.
51. McKenzie, J. C. and Mumford, P. (1965) *World Review of Nutrition and Dietetics*, **5**, 21–31.
52. Richardson, D. P. (1987) In: *Food Technology International: Europe.* The *International Review of the European Food and Drink Processing Industries*, A. Turner (Ed.), Sterling Publications Ltd, London.
53. Richardson, D. P. (1987) In: *Food Acceptance and Nutrition*, J. Solms, D. A. Booth, R. M. Pangborn and O. Raunhardt (Eds), Academic Press, London.
54. Food Advisory Committee (1987) Guidelines on the use of natural and similar terms in labelling, advertising and presentation of foods: consultative statement.
55. Codex Alimentarius Commission (1987) 17th Session, Rome 1987. *Report of the 19th Session of Codex Committee on Food Labelling*, Ottawa, 1987. Alinorm 87/22 and Alinorm 85/22A Appendix IX. Working Paper on Negative Claims.
56. Ministry of Agriculture, Fisheries and Food (1987). *The use of the word 'natural' and its derivatives in the labelling, advertising and presentation of food.* Report of a survey by the Local Authorities Co-ordinating body on Trading Standards. HMSO, London.

57. Committee on Toxicology (CoT) (1987) *Report on the Review of Colouring Matter in Food Regulations* (FdAc/REP/4/1987).
58. Federal Register: Food Labelling; Public health messages on food labels and labelling. Proposed rules Vol. 52, No. 149, August 1987.
59. The Coronary Prevention Group (1987) *Nutrition Labelling of Foods: a Rational Approach to Banding.* A consultative document prepared by the Nutrition Advisory Committee of the Coronary Prevention Group.
60. US Food and Drugs Administration (1986) *Code of Federal Regulations,* Chapter 1. Sodium Descriptors in the USA.
61. Daly, P. A. (1976) *J. Consumer Affairs,* **10**, 170–8.
62. Jacoby, J., Cheshunt, R. W. and Silverman, W. (1977) *J. Consumer Research,* **4**, 119–28.
63. Rusoff, I. I. (1978) *Food Technology,* **32**, 32–6.
64. Schrayer, D. W. (1978) *Food Technology,* **32**, 42–5.
65. Klopp, P. and MacDonald, M. (1981) *J. Consumer Affairs,* **15**, 301–16.
66. Freiden, J. B. (1981) *J. Consumer Affairs,* **15**, 106–44.
67. Soriano, E. and Dozier, D. M. (1978) *J. Applied Nutrition,* **30**, 56–65.
68. Jeffrey, R. W., Pirie, P. L., Rosenthal, B. S., Gerber, W. M. and Murray, D. S. (1982) *J. Behavioural Medicine,* **5**, 189–200.
69. Olson, C. M., Bisogni, C. A. and Thomey, P. E. (1982) *J. Nutrition Education,* **14**, 141–5.
70. National Heart, Lung and Blood Institute (1985) *Foods for Health: Report of the Pilot Program.* A pilot nutrition education program with Giant Food Inc. in cooperation with the National Heart, Lung and Blood Institute. NIH Publication No. 85–2036, reprinted March 1985.
71. Freckleton, A. M. (1985) *Qualitative Evaluation of the Tesco Healthy Eating Programme,* Food Policy Research Unit report to Tesco, Horton Publishing Ltd, Bradford.
72. Freckleton, A. M. (1987) Nutrition labelling — the role of government, the food industry, consumer advocates and consumers in the development of nutrition labelling. PhD thesis, University of Bradford.
73. Ministry of Agriculture, Fisheries and Food (1980) *Second Report of the Food Standards Committee on Claims and Misleading Descriptions* (FSC/REP/71) HMSO, London.
74. Freckleton, A. M. (1987) *Nutrition Labelling: an International View,* Food Policy Research Unit report, Horton Publishing Ltd, Bradford.

Report of Discussion

Rapporteur: MARGARET ASHWELL

The British Nutrition Foundation, London, UK

CHAIRMAN'S INTRODUCTORY NOTE

The presentation of the conclusions of the Working Group on Public Perception and Understanding elicited extensive and lively discussion which was led by Dr J. Verner Wheelock, Director of the University of Bradford Food Policy Research Unit and Dr Richard Cottrell, Science Director of the British Nutrition Foundation.

What is recorded here is not a verbatim account of the discussion in which discussants are identified, but the Rapporteur's summary of the main issues that attracted interest in the discussion.

Dr Wheelock opened by drawing attention to the increasingly prominent role and power of the supermarkets, in the hands of a small number of large retailers. As well as giving an ever wider range of choice to consumers, the supermarkets were also outstanding for the manner in which it was now possible for them to respond to rapid switches in purchasing patterns. Dr Wheelock's second main point concerned the role of the scientists. They had to face responsibilities that perhaps they had not confronted before. Although it was important that they retain their scientific integrity and capacity for critical appraisal uninfluenced by commercial pressures, there were issues that were of national importance on which it was their duty to give positive advice. There were matters concerning the appropriate dietary intake consistent with good health which the public wanted answers to now — they could not wait twenty years for scientists to dot all the 'i's and cross all the 't's. Scientists should be prepared to make positive statements.

Dr Cottrell responded by suggesting that nutritional issues were

59

rarely so clear-cut that scientists could make the sort of positive statements that might be demanded by those who were anxious to make progress in public nutrition policy, without sacrificing integrity. If issues were not clear-cut, it was the duty of scientists to say so. The public would have to accept that clear statements often could not be made. Dr Cottrell's second main point concerned the nature of the food business. It was a high risk business, subject to numerous pressures and changes. We had to ask whether far reaching changes, which might affect the viability of the industry, and arising in response to rapidly changing fashions, were in the long term interest of consumers.

The following distillation of the ensuing discussion is divided into three main sections based on the major themes of the Working Group's paper: The Medium, The Message and The Motivation.

THE MEDIUM

All too often, scientists feel frustrated that the media, wilfully or not, misunderstand their carefully weighed views and print material that is, at best, regarded as scientifically ill-considered.

Scientists should understand that the constraints placed upon the media are quite different from those that they themselves contend with. For journalists, deadlines are all important and always shorter than those of the scientist. The media can be slaves to advertising, which puts extra restraints on editorial impartiality especially when financial considerations come into force. In journalism, filling space can often be more important than getting facts right.

The media do not see themselves as conduits for the views of scientists or anyone else. They respond to the public's desire to know the answers to questions quickly. Scientists are often unprepared to do this, to admit that they do not know or that no answers are possible. The scientific community should be prepared to speak out, even on the basis of incomplete evidence: in contrast, the businessman expects to make important decisions on incomplete evidence all the time.

Scientists too have their problems and constraints. They often do not agree amongst themselves, either about the interpretation of scientific data or about what constitutes good or bad media. More interaction between scientists about the need to interpret their work for the general public and how to do so would help to alleviate these problems. Even if a journalist gives an accurate account of a scientist's views, misinformation can still arise due to overzealous sub-editing, sensationalized headlines or coverlines, or even over-dramatized public-relations releases. One encouraging trend in journalistic coverage of nutrition is the realization by journalists that, sometimes, the consumer is just as interested in knowing the 'grey' rather than being told a simplistic 'black and white' story.

A development that could help considerably in the longer term is the Media Resource Service, recently set up by the CIBA Foundation. This acts as a buffer between journalists and scientists and ensures that journalists can talk quickly to those scientists willing to enter into a dialogue, at times that are convenient to them. Nevertheless, the system still cannot totally prevent the dissemination of misinformation since journalists do not always check back their facts with the scientist.

THE MESSAGE

There was general agreement that 'the nutritional message' seems to have a high profile in the UK now. The main reasons are probably: the impact of the publication of the 'NACNE' and 'COMA' reports; the greater interest by the UK media, whether or not the information is accurate; and the sensitivity that any country has when it is near the top of the league for deaths due to CHD and when a significant number of medical authorities are claiming that diet is an important contributory factor. (This sensitivity has previously been shown by the USA and Finland: their death rates from CHD have certainly dropped but nobody is sure how much the adoption of 'the nutritional message' has contributed to this.)

Nutritional messages are some of the most difficult to understand because they are concerned with the whole body, the whole time and because eating is something that everyone must do to live. In contrast, the other lessons learnt from medical science are much easier to state and to understand, because they usually only apply to one part of the body (e.g. clean your teeth regularly; have your eyes tested regularly).

Whether 'the message' is accurate or not, the increasing power of the big supermarket chains allows the propagation of the message and the range of choice and frequency of buying food allows rapid and/or large responses to 'new messages'. Because of this potential impact on significant numbers of people, it is important to get the messages right and this is where the interaction of scientists with marketing people is so important (see above: 'The Medium'). There has been an attempt to spread the messages of 'NACNE' and 'COMA' to a mass market via the Joint Advisory Committee on Nutrition Education (JACNE) but this is generally recognized as not having been a success. The reasons are not entirely clear but there is a general feeling that the exercise was not 'marketed' properly and that it relied too heavily on self-assessment.

THE MOTIVATION

The Food Industry needs to respond very positively and rapidly to 'the nutritional message' if it wants to develop a competitive edge in marketing products. The Industry's 'free-for-all' on making claims at the moment could be seen as misleading the consumer, but could also be seen as providing the motivation in terms of stimulating him or her to be more aware of nutrition.

Nutrition labelling is not the same thing as nutrition education; for example, if macronutrients are expressed in terms of percentage of energy in a product, the consumer still needs to relate this to total energy in the diet. He will not get this information from the label on a can of baked beans!

Manufacturers' and retailers' interests do not always coincide. Some retailers, for example, responding to perceived consumer concerns, demand that manufacturers remove all additives. Applied indiscriminately, this may not be in the long term interests of either manufacturers or consumers.

Clearly, diverse but complementary methods of nutrition education are required. In The Netherlands, there are regional Nutrition Advice Bureaux. These give consumers easy access to information; but there is no evidence yet that this has changed dietary habits or brought down the rate of CHD.

Easy access to extensive information might not be the only answer to motivation. Everyone has different problems, needs and tastes. Greater availability of a wide range of products is as important. The South-East is particularly well served by retailers in this respect but other areas of the country may not be so fortunate.

It is not feasible for government to manipulate the food supply to promote health. We have not, after all, seen the government taking steps to ban smoking completely, even though the health risks associated with it are much greater.

Scientists differ widely about whether they should be involved in policy decisions of this sort. Some agree with Winston Churchill that they should be 'on tap' but not 'on top'. That is to say, they should participate in advisory committees and no more. Others strongly disagree with this approach and act accordingly.

Finally conflicting views were expressed about the degree to which the agricultural industry could or should respond to perceived needs for changes in the composition of food raw materials, for example by producing leaner carcasses or breeding for cattle to produce lower fat milk. It was suggested that the agricultural industry should examine the situation in the USA where the CHD rate has declined significantly over the last twenty years. It was pointed out, however, that the reasons for this drop were not at all clear and were likely to be more complex than simply a result of changes in diet. Indeed, the average consumption of total fat and of saturated fatty acids had changed very little in the USA.

Again, economics, rather than fashions in nutrition was likely to be the dominating factor. Farmers were unlikely to respond to demands for changes in the composition of agricultural produce unless there were clearly perceived benefits in terms of profitability. Once more the retailers would probably have the dominating influence in pointing out new opportunities for agricultural production.

Appendix; Report on Questionnaire on Food and Health Information

B. A. ROLLS[a], A. F. WALKER[b], and T. E. SWINDELL[c]

[a]AFRC Institute of Food Research, Reading, UK
[b]Department of Food Science and Technology, University of Reading, UK
[c]Glaxo Group Research, Greenford, UK

As we have indicated elsewhere, many people working in nutrition have reservations about the material that appears in newspapers and magazines on this subject. We have devised a questionnaire designed to throw some light on the way that nutrition information reaches the public.

It was sent to editors of newspapers and magazines, those responsible for choosing the material placed before the public, and sought to elucidate their own views on nutrition, the knowledge they might have, how they would choose features or articles and from where they would seek advice or information. (For reasons of time and expense we restricted ourselves to the printed word. Important as they undoubtedly are, radio and television programmes were excluded.)

Perhaps arbitrarily, we divide these publications into three: the general, those that deal with a range of interests (newspapers, 'women's magazines' and so on), specialists, those with a particular interest in food or nutrition (slimmer's magazines, catering journals) and professionals, those read by people with a professional concern in this area (those for dietitians, nurses, medical practitioners — from whom many people will seek advice).

Of 135 questionnaires, 58 were returned (44%) of which 12 were eliminated or eliminated themselves; 42% general, 27% specialist, 31% professional. (This does not reflect those sent out; it indicates the greater inclination of those with a particular concern with the field to reply.) Of the editors, 27% had some formal qualification in the area of food, nutrition or health; as might be expected, editors of professional magazines were more likely to be so qualified.

The circulations varied widely, from around 500 to well over 1 000 000. Most claimed to appeal to all ages, or those in active life (between 20 and 60). More than half (52%) thought their readership was predominantly female, only 6% male; 42% considered their readers were of both sexes.

The respondents were asked whether their publications appealed to any particular section of the community; 77% thought they did. Of those who identified target groups, 'generalists' mentioned housewives (6 times) working and non-working mothers, businesswomen, working women, high income groups, feminists, left and gay feminists. 'Specialists' appealed to slimmers, the health-conscious, keep-fit and sports enthusiasts and food sector workers. 'Professionals' were overwhelmingly directed to medical or paramedical practitioners.

Respondents were invited to tell us whether features on the relationship between diet and health would be of 'very great', 'considerable', 'moderate', 'marginal' or 'little or no' interest to their readers. Whether from tact or honesty, all but one respondent marked one of the first three columns, mainly the first or second. Most publications thought that up to a quarter of the articles published lay in this general area; those with a special concern tended to have more.

We have often heard the view expressed that whereas most newspapers and magazines will give a fair and unbiased account of (say) a research report, or a new publication from one of the Royal Colleges, the more dubious material originates from writers outside the editorial staff, writers with no understanding of nutrition or with a particular eccentric view. This questionnaire provides no evidence for that view. We asked the editors to estimate the proportion of articles that were commissioned, produced by editorial or research staff or chosen from unsolicited submissions. They were also asked whether they would carry out any checks on the writer(s). Over 60% accept no unsolicited material, and about the same proportion would always check the credentials or qualifications of any writer, and this varies little with the type of publication. The problem — and we shall return to this later — is how someone necessarily unversed in the nutritional sciences can select an 'expert' from whom to commission articles and an unbiased body from whom to seek advice.

We tested the attitude of editors towards material from what we called 'special interest groups', those with a commitment to putting forward a particular point of view, such as the Butter Information Council, the Sugar Bureau, the Food and Drink Federation. Some 48% either would not use such material or would seek a second opinion first — some added comments that they would point out any bias they considered to be present. Some 33% would be prepared to use such material, the 'specialist' magazines being perhaps slightly less suspicious.

Overwhelmingly (75%) readers are always able to comment on articles, and in general the editors reported that features on the relationship between food and health resulted in about the same reaction as other subjects (56%) with fewer reporting less or more reaction (15% each) or much more or less reaction (5% each). If a feature provoked particular controversy or interest, most would be prepared to run an expanded letter column, less rarely to present an article giving an opposing or alternative view. Some pointed out that the time-scales involved in publication made this difficult.

The respondents were asked what influenced their choice of features and writers. (They were asked to give various factors a score between 0 — unimportant — and 5 — very important, so the higher the mean score the more weight given to such considerations.) In choosing features, most weight is given to new or interesting ideas (3·9) and a sound factual basis (3·7), rather less to clearing up confusion in readers' minds (3·0), challenging or controversial material (2·8), a practical slant (2·4) or visual appeal (1·8). Some gave other reasons, including topicality and pure chance.

When choosing writers, most claimed to be influenced by scientific or medical respectability (2·8), a good writing style (2·6) and personal contact (2·3). Less weight was given to articles appearing elsewhere (1·8) or the appearance of an individual in a news item (0·7). Several added that reliability and the ability to keep to publishing deadlines would affect their choice.

The respondents were asked to score in a similar way different aspects of the relationships between diet and health according to how important they thought their readers considered them. Food additives (3·2) and dietary imbalance (3·1) — that is a poor choice of food — came well ahead of naturally occurring harmful substances, pesticide residues and environmental contamination (1·9). Microbiological contamination, the most important source of food-borne ill-health, was well down (1·7), barely scraping in ahead of industrial contamination (1·4). Our respondents told us that we should have been concerned with other subjects, notably social aspects of food and food intolerances and allergies. Other topics mentioned were very-low-energy and crash diets, vegetarianism, irradiation, over-processing, hormone and antibiotic residues in meat and 'the relationship between diet and performance', a phrase that we hesitate to interpret.

We then went on to enquire about the respondents' own views on food and health, and began by asking them to indicate how important a question this was. Most (58%) were prepared to consider this of considerable importance and 19% went so far as to call it one of the most important questions of today. Some 10% thought the diet and health question of moderate importance, while 8% thought there were really other more important matters. Some 4% ticked the final box, dismissing the debate as something of little importance that has been much exaggerated. There was little difference between types of publications.

The next three questions asked whether they had heard of influential reports in the field of nutrition and health, various prominent writers in the area and organisations active in this work. Some names were supplied, and they were invited to add any others they wished. Several respondents told us that there were too many to mention, an obvious problem for specialists; however, we were more interested in the impact of these on the general editor.

About three-quarters of the respondents had heard of the NACNE (77%) and COMA (75%) reports; JACNE (58%) and the National Food Survey (56%) made less impression. About a dozen other reports and books were mentioned, none more than once.

When questioned about the three prominent nutrition scientists and two authors of popular nutrition books and articles, the editors revealed that 73% had heard of Arnold Bender and Geoffrey Cannon, rather fewer of Maurice Hanssen (60%), Philip James (48%) and John Garrow (35%). Asked to supply other names, our respondents provided a list of 32, of whom only seven appeared more than once. Don Naismith, Dorothy Francis, Miriam Stoppard, Leslie Keaton and Erik Millstone appeared twice and John Yudkin three times, but the runaway 'write-in' winner was the late Caroline Walker, mentioned by no fewer than seven people.

Of the organisations, the HEA (98%) and MAFF (96%) were most well known, with creditable showings by the BNF (83%), the London Food Commission (73%) and the BDA (69%). We were surprised by the low rating of the Food Additives Campaign Team (42%), in view of our impression of its high profile. A very wide range of other organisations was given (total 18), none more than once, by 31% of those replying.

We asked whether the respondents had approached any organisations during the past year for factual information in a specialist area or to comment on a feature they were considering. We were trying, not too subtly, to find whether the editors made any checks on their material, and whom they would be inclined to approach. A high proportion had done so (65% for information, 45% for comments) from a wide variety of sources. Words along the lines of 'too many to list' appeared often, but some 24 individuals and organisations were mentioned.

A number of publications have 'in-house' experts trained in nutrition, dietetics or medicine; not only will these individuals know the answers to many questions, they are also likely to be aware of whom to consult about the remainder. A number referred to the Royal College, GPs and other individual medically-trained individuals and various university departments or academics. The organisations specifically named more than once were: MAFF (7), HEA and London Food Commission (5 each), BNF and BDA (4 each), DHSS and FACT (3 times each) and the Soil Association (twice).

A problem that is by no means trivial for the editor (or any other individual) lacking specialised training is where to find advice. Is any particular organisation giving or promulgating nutrition advice unbiased, non-political, non-commercial and independent, or is its advice given from a particular stance, and not necessarily a nutritional one? This problem was addressed cogently by Neil Donnelly in BDA Advisor of Winter 1986/7, in an article entitled: *'Consumer Beware: Nutritional Advice Can Damage Your Health'*.

We left space at the end of the questionnaire to add further remarks, and it was this problem that exercised the minds of editors most — or at least elicited the most comment. One magazine, with its own 'in-house' experts took a confident line:

Edited copy is sent to manufacturers and the relevant authorities before going to press . . . but the text will not be changed without good reason.

Others were aware that whatever they did they were unlikely to get it right every time:

All articles are checked by a GP or a member of one of the Royal Colleges (or similar) because I feel we have a responsibility to our readers. This system may not be foolproof as obviously even experts have an axe to grind, but we do our best.

Such attitudes are a long way from the casual line taken by sections of the national press (and which tarnishes by association the responsible areas of the media). At the same time, the food question is of particular interest to many of our respondents, whereas in a newspaper it must compete with the unemployment figures and the latest terrorist outrage.

Our final question was which of the aspects of food and health detailed earlier they considered important. (As in the earlier questions, they could indicate any score from 0 (unimportant) to 5 (very important), and the figures given are the means.) Whether it is the result of publicity or the feeling that this is one area within the control of the individual, dietary imbalance (or the actual choice of food) received the highest rating (4·3). Little separated food additives (3·4), microbiological contamination (3·2), environmental contamination (3·1) and pesticide residues (2·9). They seemed less concerned about naturally occurring toxic substances and industrial contamination (both 2·5). If this is not the order we should have chosen (and we may be wrong!) it is closer to it than the order they think concerns their readers.

Perhaps the most revealing parts of the completed questionnaire were the final comments. Naturally, we were pleased when they echoed points we had already made:

> Changing the diet of the population for one reason could create other problems.

> I believe that tempting people to eat a better diet . . . is more effective than frightening them with shock stories.

Many were anxious to promote a measured approach:

> It is balance that is the most important.

> Too much hysteria in the media, too little informed and rational discussion.

> We prefer a commonsense approach, with moderation rather than extreme slavishness to food fads and questionable research results.

Some took a sour view of what they regarded as exploitation:

> The well-meaning nuts probably do more harm than the villains! But it's hard for a lay public to distinguish them.

> Ill-informed material from ———— has caused much harm.

It's encouraging that people are much more concerned with the effect of food on their health — but this does make them vulnerable of course, to 'experts' who don't know what they are talking about (although they may be entirely sincere and well-meaning) and to get-rich-quick villains.

And some wished for a more positive response from qualified nutrition workers:

> I wish professionals in this area would be more outspoken when self-proclaimed experts in nutrition are featured in the press or on television despite their total lack of genuine expertise. Nutritionists should be less absorbed in their own small area.

The message is that those within the field (and probably without) would like a clear indication of where they might go for unbiased advice, which returns us to the question of a formal qualification in nutrition science.

It is clear that some people hang nutrition policy upon their own political or philosophical stance. One told us we should have investigated the 'food industry lobby', saying:

> I am concerned about the manipulation of our tastes and demands through advertising by the food industry.

While another chided us for not going into 'the political relationships between the various health lobbies and other pressure groups' adding:

> An important area is the use of the food question as left-wing propaganda — the use of food policy as a political weapon.

Such posturing and in-fighting may well lead outsiders to throw up their hands and abandon the whole question:

> I have survived 60 years on the wrong diet, poor eating habits, drinking to excess and smoking 80 cigarettes a day. I firmly believe that many food manufacturers, realising that some of the public were listening to the babblings of certain faddists, saw an opportunity to jump on a bandwagon — and increase prices unjustifiably!

Truth and balance are the first, but not necessarily the only, casualties of studied misinformation.

RESERVATIONS

How valuable is this survey as an indication of how editors actually behave? Those of us who carried out the survey are well aware of its limitations. Editors of professional journals (who have a greater interest in the field, and perhaps more rigid standards of accuracy) are more likely to reply, as are editors, of whatever type of periodical, who have a justifiable pride in their own good practice. (We have had no replies from magazines in whose pages some of the most striking misinformation has appeared.)

Moreover, everyone likes to show himself or herself in the best light, and may claim as a regular practice what they would like to do under ideal conditions. It is perhaps not appropriate for academics who have never been subjected to the tyranny of the deadline to demonstrate an unsympathetic disapprobation for some media practice.

Nonetheless this survey represents, to our knowledge, the only information available on the way nutrition is presented to the public.

THEME 2

Implications for Education

Introduction

J. G. W. JONES

Department of Agriculture, University of Reading, UK

This Working Group concerned itself with formal education and regarded its task as leading on from the previous Working Group's deliberations on 'Public Perception and Understanding'. Our deliberations were based on the premise that no amount of information regarding food and its production and utilization can have any appreciable effect on the consumer or participants in other sectors of the food chain without a basis of formal education.

The group accordingly set about its task by attempting to define special areas of need with regard to formal education. The areas included:

(1) food production
(2) food processing and distribution
(3) technology transfer and safety of food
(4) catering, including the social services
(5) education of the school child
(6) infant nutrition
(7) geriatric nutrition

Contributors were commissioned to prepare papers designed to cover most of these areas. The authors were asked to review the current state of education in these areas, to point out the deficiencies which were apparent to them and to suggest what changes may be desirable; these requests have been largely met in the papers produced. The group was unsuccessful in one major area, namely that of geriatric nutrition about which there is likely to be an ever increasing need for formal education as the demographic profile of the UK changes.

There is a tendency to suppose that education is required only by

J. G. W. Jones

those not involved in the food chain except as consumers. One would be quite mistaken to hold this view. There would be no need for this symposium were it not for the fact that operators in particular sectors of the food chain are ignorant of other sectors in the chain. This is a continuing theme of the conference and I hope the outcome of this Working Group's activities has covered the formal educational aspects of this theme.

The lack of instruction upon nutrition in schools, which was noted by the Working Group on 'Public Perception and Understanding', has been addressed by our Working Group which went on further to consider provision in the schools for teaching in relation to the whole food chain. This after all is where the education of the consumer has to begin.

Our purpose then is a fairly well defined one and the extent to which it has been achieved may be measured by the papers which follow.

Education and Training in Agriculture

J. B. DENT

Edinburgh School of Agriculture, University of Edinburgh, UK

ABSTRACT

A review is made of the current structure of university courses in agriculture, both in Great Britain and overseas, and including the CNAA Degrees in agriculture available within the United Kingdom. Attention is drawn to the changing needs of the rural sector in terms of employment. Employers of graduates do not always need highly specific skills, but ability in communication, computer literacy and numeracy are frequently as important as disciplinary knowledge.

Education opportunities in agriculture at National Certificate, National Diploma and Higher National Diploma levels are presented. Reference is made to the fact that many students who qualify at Higher National or Diploma level seek careers in similar areas to about 60% of university graduates in agriculture.

Skills training in agriculture in the United Kingdom are offered by the Agricultural Training Board and through the National Farm Craft Proficiency Test Committee. The Agricultural Training Board provides appropriate training to new entrants, to established workers, as well as to management personnel.

The provision of postgraduate education in agriculture within the university framework is discussed. Mention is made of the fact that many courses at Masters level in agriculture are attractive to students from overseas, as are opportunities for research study at PhD level.

It is concluded that many current educational courses in agriculture at all levels stop too sharply at the farm gate. There is a need to recognise the powerful interaction between the rural debate on the one hand and the demand for food as expressed in buying habits of the public on the other.

EDUCATION

Degree Courses

There is a surprising similarity between degree courses in agriculture science offered at many university centres. Tables 1 and 2 illustrate the broad course structure at the University of Nottingham and the University of Edinburgh. These courses are characterised by a first year which concentrates on advancing science understanding and limited choice beyond a general specification of interests which mainly occurs in the final year of a course. Agricultural degree courses offered at the University of Reading and Wye College at the University of London (Tables 3 and 4) follow broadly similar patterns, although these courses do offer a certain degree of industrial context setting in the first year of the course. Without this element, students of agriculture frequently are not stimulated in an academic sense and find the first year somewhat frustrating. The features of minimum choice and lack of opportunity for students to stamp individuality on their studies, appears evident in all examples. The major redeeming feature offered at all the universities cited, is the opportunity for all honours students to complete a dissertation in their final year which creates the opportunity for individual study and flair to be expressed.

The NCAA degree in Agricultural Technology at Harper Adams and similar qualifications offered at other agricultural colleges, offer the student a more applied approach. Table 5 illustrates the framework for

TABLE 1
Degree Course Structure, University of Nottingham, BSc(AgSci)

Year 1	Year 2	Year 3
	Agriculture	3 subjects from limited range
	Horticulture	3 subjects from limited range
Biochemistry	Plant scient	3 subjects from limited range
Biology		
Physics	Animal science	3 subjects from limited range
	Food science	3 subjects from limited range
	Environmental science	3 subjects from limited range

TABLE 2(a)
Degree Course Structure, Edinburgh (Agriculture)

Year 1	Year 2	Year 3	Year 4 (Honours)
Chemistry	Environmental biology	Machine management	Enterprise planning Agricultural marketing
Biology	Agriculture 2	Crop production	Farm analysis and planning
Industrial management	Resource economics 2	Animal production	Systems and methods
Agricultural systems	Soil and water 2	Rural resource management	Communication and extension + 4 electives

TABLE 2(b)
Degree Course Structure, Edinburgh (Agricultural Science)

Year 1	Year 2	Year 3	Year 4
Chemistry		Animal science	Animal science
Biology Physics/maths	Four elective science units	or	or
Agricultural systems		Crop science	Crop science

TABLE 3
Degree Course Structure in Agriculture, University of Reading

Year 1	Year 2	Year 3
Plant science (2) Physics (2) Biochemistry/nutrition (1) Economics (1) Agriculture (2)	Crop production (1) Animal production (2) Engineering (1) Maths (1) Management/economics (2)	Animal production Crop production Engineering Buildings Management systems

TABLE 4

Degree Course Structure in Agriculture, Wye College

	Year 1	Year 2	Year 3
Agriculture: field	Agriculture Livestock production Anatomy/physiology Crop botany Soils Economics 1	Economics of agriculture Management 1 Animal production 1 Crop production 1	Animal production 2 Crop production 2 Farm mechanics Management 2
Agricultural economics: field	Economics 1 Maths Statistics 1 Livestock production Agricultural technology	Economics of agriculture Economics 2 Economics 3 Statistics 2 Management 1 Marketing 1	Economics 4 European agricultural policy Special study
Business management: field	Management Economics 1 Computing Agriculture 1 Mechanisation Livestock production	Management 1 Economics of agriculture Marketing 1 Land/estate management Animal production 1 Crop production 1	Animal production 2 Crop production 2 Farm mechanics Management 2 Management 3 Management in practice Marketing 1

NB. (1) At least 6 subjects must be passed in each year block.
(2) Other fields of study — Horticulture; Applied plant science; Animal science.

TABLE 5
Degree Course Structure, Harper Adams BSc (Agricultural Technology)

Years 1 and 2		Year 3	Year 4
Biochemistry	Crop production technology	Practical experience	Business management
Microbiology			Decision making
Genetics			Management of agricultural staff
Breeding	Animal production technology		Agricultural policy
Agricultural biology system			Investment projects
Agricultural economics system			Review projects
Farm business system			Economics in agriculture
Agricultural marketing system			Assignments in technology

the Harper Adams BSc (Agricultural Technology). The subjects within the degree course are of a different complexion to those offered at the universities. Relevance is inserted into the first year of the course and some level of specialisation is offered between animal and crop production right from the beginning of the course. The further major distinguishing feature is the true 'sandwich' nature of the course which allows the student an industry training year after two years of study: a year of applied and structured training. The final year is also clearly differentiated from that of the universities in that while the opportunity is offered for students to work on their own and develop their own interests, it is a synthesising year rather than a year of specialisation. No doubt, many students find this format attractive and challenging.

An example of a more recently designed degree is the Bachelor of Commerce (Agriculture) offered at the Lincoln Agricultural University College in New Zealand. The format for this degree is shown in Table 6. This degree is characterised by the insertion of the 'professional' subjects offered early in the degree structure in order to capture interest and provide background for application for more conceptual elements of the course which follow. The degree offers a wide range of subject choice for students, but also several broad fields of study, examples of which are given in Table 6. All the fields of study demand practical industry experience and the Farm Management field of study requires 72 weeks of farm experience prior to graduation. This course has proved extremely popular. It should be further noted that while students are recommended to operate within a selected field of study, individuals can elect subjects to make up their degree in any way they find attractive, limited only by prerequisite requirements for advanced subjects.

The above degree structures should be seen in the light of the historical employment statistics for graduates from agriculture and agricultural science courses in the universities. Figure 1, for example, provides a breakdown of the main employment categories of students graduating from the University of Edinburgh over the years 1982–6. Over 30% of students became involved in farming operations either in the UK or overseas so that in total over 50% became closely involved in agricultural production in some form or other. A further large group of students accounting for approximately 20% of the total became involved in either further study or research in some capacity. A further large proportion found employment completely outside agriculture.

The emphasis in agriculture is changing quite dramatically in the face of economic and political adjustment. Some of the issues which are

TABLE 6
Lincoln University College of Agriculture, BCom(Ag) Course Structure

Year 1	Year 2[a]	Year 3
Accounting	Select 8 units	Select 8 units
Computing	from 32 offered	from 32 offered
Economics 1		
Maths		
Statistics		
3 Electives		

Fields of Study: Farm accounting; Farm management; Agricultural marketing; Economics; Econometrics; Agricultural business management; Applied computing; Rural valuation; Urban valuation; Financial management.

Example 1: Accounting: required for professional recognition

Year 2	Year 3
Financial accounting	Accounting theory
Management accounting	Auditing
Accounting information system	Farm and trust accounting
Financial management 1	Tax law
Business law	Accounting law
3 Electives	3 Electives

Example 2: Farm Management

Year 2	Year 3
Livestock production	Advanced farm management 1
Plant production	Advanced farm management 2
Farm management 1	Advanced farm management analysis 1
Farm management analysis	Advanced farm management analysis 2
Management economics	4 Electives
3 Electives	

[a]NB. (1) Units in Year 2 depend on those required in Year 3 and may act as prerequisites.
(2) Up to 72 weeks of industrial experience is required.

facing professionals in agriculture now and in the immediate next few years relate to alternative land use (diversification of farm enterprises, conservation issues, recreation possibilities within the countryside, forestry development and an emphasis on countryside matters rather than on farming), reduced capital inputs, more emphasis on market definition and supply, lower farm trading margins, problems with debt

J. B. Dent

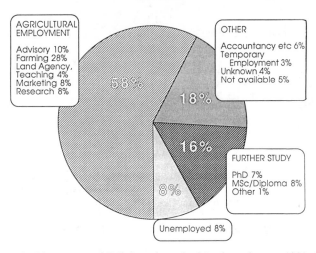

AGRICULTURAL EMPLOYMENT

Advisory 10%
Farming 28%
Land Agency,
Teaching 4%
Marketing 8%
Research 8%

OTHER

Accountancy etc 6%
Temporary
  Employment 3%
Unknown 4%
Not available 5%

FURTHER STUDY

PhD 7%
MSc/Diploma 8%
Other 1%

Unemployed 8%

FIG. 1. Placement of Edinburgh agricultural graduates, 1982–6.

servicing, a changing structure for advisory services to agriculture and some farmers facing major financial difficulties. The question is, are our current graduates going to be equipped sufficiently well to cope with such problems, and will they be able to adjust during their career lifetime to further changes to the industry? Most graduates will make their valuable contribution but university courses should reflect not only the change that has already taken place, but prospective changes in industry in order to give graduates the best opportunity and to provide the best agents for change and adjustment for the industry.

The problem is then for those people concerned with the design of courses to understand the change in demands that are likely to be placed on graduates from UK agricultural university departments over the next period of years. Although it would appear to be sensible for such people to be in close contact with the employers of agricultural graduates, experience has shown that such employers generally take the view that university professionals know their job and are putting together courses that are relevant and challenging to students. Within limits, it seems that the exact subject matter read is of little concern and while employers wish to keep in contact with university staff in the broad orientation of courses, by and large they do not wish to direct change. Certainly, when employers are interviewing for new recruits they are usually looking for the 'right' person, as well as for someone with good and/or relevant academic qualifications. The characteristics that employers list as important seem to include: a sharp, well motivated person who can take initiatives; a person who can work in a team and be flexible and adaptable; a discernible level of ambition; a person who is a good communicator and who can quickly and effectively assimilate data and can pass information on to colleagues and superiors. These concepts accord well with the statement made by Sir David Smith, Principal of the University of Edinburgh,[1] who noted that industrialists are not necessarily looking for graduates who are highly trained in a specific craft, but rather for 'intelligent people in whatever subject area, who are articulate, can think, take responsibility, have initiative, stand up and talk and present things well'. Similar concepts have been expressed by Bell[2] as past Chairman of the Association of Graduate Recruiters, who indicated:

> Employer preferences are based not so much on the subject as on the applicant's possession of underlying commercial skills such as numeracy, effective communication, computer knowledge and an understanding of business. Many of the skills essential for a fruitful career in industry or commerce can be found in any discipline, but that does not mean that we will take a graduate with any degree subject regardless of other factors.

The current Chairman of the Association of Graduate Recruiters, Perkins[3] is also of the view that 'the ideal for most employers is to match specific posts to particular combinations of degree subjects, but a market place where demand for good graduates greatly exceeds supply is forcing employers to take a wider view'.

University departments of agriculture will remain under pressure in the foreseeable future: pressure created by financial limits that will demand more efficient teaching and probably higher student : staff ratios; and pressure to ensure that the quality of students entering degree courses in agriculture does not continue to fall. The latter implies that the Departments of Agriculture in the universities must be competitive for students *vis-à-vis* other disciplines, as well as in relation to other non-university degree awarding institutions. Perhaps university degrees in agriculture are at a crossroads and clear minded decisions are needed about the role of university education in agriculture just as strong and clear minded decisions are currently being contemplated about agricultural education at HND and CNAA degree levels.

Diploma Courses

Higher National Diploma
Fifteen colleges in the UK offer courses leading to a Higher National Diploma (HND) in agriculture and allied subjects. They are structured on a 'sandwich' basis in which the second year of a three year programme is a structured learning experience in industry. The HND courses are precisely targeted to meet clear objectives and the aim is to produce technologists who have an education relevant to the present and future needs of farmers and farm managers and those seeking employment in related industries. The course is essentially based on a good understanding of the principles of science, economics and agricultural practice and it enables students to acquire appropriate knowledge, skills and personal qualities which will assist them to respond to changing circumstances in the future. The HND offers an impressive curriculum which integrates the basic components of agriculture in strict format. Students who successfully complete the HND in agriculture look for jobs as technical advisers, field officers and technical sales representatives, in teaching and lecturing in county agricultural colleges, as enterprise managers, farm managers or farming in their own right. If these career expectations are compared with those set out for Edinburgh agricultural graduates in Fig. 1, a considerable overlap can be determined. About 60% of the graduates would appear to be aiming for the same general career areas as those graduating with HNDs and are therefore likely to be in direct competition with the diplomates as well as graduates from the CNAA degree programmes. For these sorts of careers, the balance of favour may not always fall with the agricultural graduate from the university who generally has less practical experience, less orientation towards the industry and whose industry skills are likely to be substantially less well developed.

Other HND courses are offered in various special areas at particular colleges throughout the country: in 'agri-business'; in agricultural marketing and business administration; in agricultural engineering; in rural resources and their management; in applied biology; in milk technology and advanced food technology; and in poultry husbandry. Some colleges offer post-HND courses again in rather special areas, for example: Harper Adams College offers a post-HND course in crop protection; the Royal Agricultural College and Seale Hayne College offer post-HND courses in advanced farm management; and the Welsh Agricultural College offers post-HND courses in beef and sheep production and marketing. Students who undertake these advanced courses are targeting their future careers quite precisely and have high capability in the areas of specialisation which they follow.

National Diploma

The National Diploma in agriculture is offered widely throughout the country at over 30 colleges, some providing particular biases: some six colleges offer courses leading to diplomas in food technology (dairying); two colleges providing programmes leading to diplomas in poultry production; 19 colleges offer specialist diploma courses in subjects including farm secretarial work, forestry, arboriculture, agricultural engineering, agricultural merchanting and in countryside recreation.[4] Most of the courses are organised on a three-year sandwich basis, similar to the HND with the second year spent working in the appropriate industry. The courses offered at this level are more skills-orientated and practically-related than those offered within the HND. Students completing the qualification find jobs as skilled operators within the appropriate sector of the industry in which they have tended to specialise.

National Certificate

National Certificates in agriculture are provided within a large number of agricultural colleges with the aim of providing a basic grounding in relation to the principles and skills of agriculture, including crop and animal production, farm machinery and farm accounts and records. The one-year syllabus followed in any particular agricultural college will often be related to the farming in the local area and much of the course content is of a highly practical nature allowing students to carry out tasks in workshops and on farms. Equivalent National Certificates are available in dairying, agriculture with home economics and for farm secretarial training.

Students who successfully gain National Certificate qualifications tend to look for jobs at the craft or supervisory level in the industry, although some do achieve managerial posts or become farmers in their own right. Trained farm secretaries may gain employment on a single, large farm or may service the secretarial requirements of a number of farmers travelling from one to another at regular intervals. Following a one-year training at national certificate level, an advanced national certificate is available both in farm management and general agriculture for those who do well enough in the first year of national certificate training. Generally the advanced certificates also allow some form of specialisation: in pig unit management; farm business management; or sheep management. This level of qualification will lead to jobs, again at the skills level, but will eventually provide opportunities to take positions as unit managers after several years of work experience.

TRAINING IN AGRICULTURE

The Agricultural Training Board

Direct training services to the agricultural industry are provided under the Agricultural Training Board. This institution was established in 1966 to ensure that the industry was 'training-conscious' and to ensure that the industry was continually aware of training needs and was able to organise to meet these training needs with or without assistance. In order to fulfil these objectives, the Board relies heavily on the expertise of people in the industry: for example, from highly skilled agricultural and horticultural craftsmen, engineers from machinery manufacturers and college lecturers. The Board categorises the training provided under three main headings: new entrants to the industry; established workers in the industry; personnel in the industry concerned with management supervision and finance. Approximately 600 training groups have been established under the auspices of the Agricultural Training Board in order to enable employers to share their facilities and equipment for training purposes, thereby providing smaller businesses with the benefit of a service comparable with that of training staff of a large firm.[4]

The provision of training courses to provide basic skills and assisting in the development of the ability to carry out professional tasks within an industry as diverse as agriculture is a crucially important capability. Agriculture has and continues to benefit from the way in which the Board has played its role in upgrading the manpower capability available to the industry.

The National Farmcraft Proficiency Test Committee

This scheme, although co-ordinated at a national level, relies on area organisations to organise proficiency tests in accordance with conditions laid down by the national councils. It adopts a national scale of marks for the test and appoints area stewards who have the responsibility of ensuring that the required high standard is maintained at tests throughout the area. All candidates for the test must be members of the Association of Young Farmers' Clubs and must be guaranteed by the Club leader as having a reasonable chance of qualifying. The training programme for the tests is based on the Agricultural Training Board's Craft Training Schedule. In order to pass a test in a farm craft, a candidate irrespective of age, must have reached the adult standard that a good farmer would expect of a worker. Certificates are provided in a wide range of farm crafts including, for example: tractor driving; sheep shearing; fertiliser distribution; ploughing; fencing; and operating harvesting machinery.[5]

POSTGRADUATE EDUCATION

Postgraduate training in agriculture is provided by most university departments of agriculture and related disciplines. Courses at Masters degree level, usually of 12 months duration, are offered in order to take the graduate's knowledge and information beyond that offered at first degree level. In most cases, a full academic year of course work is followed by an independent study in a selected speciality which will be presented as a dissertation. Occasionally these courses are taken by graduates in disciplines other than agriculture in order to redirect their education so that they can contribute specifically to the agricultural industry. In other cases, opportunity is taken to provide agricultural graduates with specialised areas of training which extend their knowledge and experience. Often these courses are attractive to graduates in agriculture from countries other than the UK. A large number of students from Third World countries will be sponsored by their national governments to undertake Masters training in an agricultural department of a university in the UK. Usually the courses will be of a particularly specialised nature such as seed technology, business management, agricultural engineering, animal physiology, etc. The objective of students taking these courses is usually to make them more competitive in their search for employment in specific areas of the industry.

Programmes leading to a research degree at PhD level are also provided by most university departments of agriculture. Students may be required to take a number of specific courses prior to entering into their research programme, or even to audit courses set at Masters level. Usually the programme of work is for a minimum period of three years after first graduation. Because of the specialised and narrow nature of the research involved, the students undertaking work towards this degree will usually be looking for a career in a research establishment associated with agriculture. However, it is often considered that students who engage in the detailed research and investigation involved in successfully completing a PhD have undergone a period of rigorous training which will fit them well for a range of career options in the future. However, the fact remains that most people who qualify at this level initially find themselves attracted to positions in which detailed research forms the large part of their employment. In many developing economies, it appears to be necessary for a person who is to find employment at a high level in government circles to have a PhD qualification. It is not surprising therefore that the demand for places on PhD programmes in British agricultural university departments is continuously high from overseas graduates.

CONCLUSIONS

Perhaps one of the pressing needs within the agricultural industry is for producers to understand that theirs is an increasingly small role in the whole production, distribution and marketing process for food. In a market situation which is characterised by rapidly changing demographic parameters and economic adjustment, demand for agricultural commodities is determined by the presentation and display of products which may bear relatively little similarity to the end product of the farming process. To fulfil their future role within the whole market chain, farmers must demand better information about their products as they proceed through the various distribution and processing stages to the market so that production processes can be adjusted and directed by the price mechanism.

Only specialist interest groups of students (for example, agricultural economics (agricultural marketing)) within agricultural degrees at universities and in central institutions gain some concept of the processing of farm commodities and the changing nature of the market process. However, most students in agriculture at degree and diploma levels develop relatively little understanding of this complex process. It is rather difficult to judge the extent to which such material finds its way into agricultural courses because it is largely 'buried' within larger curriculum items (for example, marketing or trade), but there is no doubt that in comparison to production technologies these matters are treated as minor elements in courses. The importance of this Conference is that it forces attention on the need to integrate between the various elements within the food production chain. The Achilles' heel of current courses in agriculture at all levels is that they tend to stop far too sharply at the farm gate. They do not underline the vital need for information flow along production, marketing and consumption chains as a means of improving economic and energetic efficiency. They do not overtly recognise the powerful interaction between the rural debate on the one hand, and the demand for food as expressed in buying habits on the other.

REFERENCES

1. Smith, D. (1987) *Edinburgh University Bulletin,* June/July, p. 14.
2. Bell, K. (1987) Graduates are for Converting, *Sunday Times*, 2 August, p. 63.
3. Perkins, H. (1987) Graduates are for Converting, *Sunday Times*, 2 August, p. 63.
4. Anderson, A. M. (1984) *Courses in Agriculture in the United Kingdom,* The Agricultural Education Association, Askham Bryan College, York.
5. Scottish Association of Young Farmers' Clubs (1986) *Proficiency Tests in Farm Craft.*

Education for the Food Industry*

A. J. YOUNGS

Department of Education and Science, Preston, UK

ABSTRACT

UK provision of initial and post-experience education in food science, technology and manufacture is considered from a historical, current and future perspective.

Changes in the food industry since the 1940s are related to the growth of broad courses in food science and technology at advanced level now offered at 25 United Kingdom institutions and the decline in courses at craft and supervisory levels. The approximate balance between the supply of potential employees from advanced level courses and industry demand is noted.

The implications for education of a range of current issues are discussed. These include curriculum content, industrial experience and student competence, commercial awareness, student backgrounds and industry liaison. The need for education to have a clear view of market needs and particularly how they should be prioritised is identified. The case for a forum at which representatives of all sectors of food education are represented is raised.

Provision of post-experience education is reviewed including the role of professional bodies, education establishments, industry, non-statutory training organisations, distance learning agencies and the Professional, Industrial and Commercial Updating Initiative. The challenge to develop flexible modes of delivery is mentioned.

*The views presented in this paper are those of the author and do not necessarily reflect those of the Department of Education and Science.

INTRODUCTION

Professor John Hawthorn, contributing to an Institute of Food Science and Technology symposium on '*Education in Food Science and Technology*' held in 1970 presented a paper entitled 'The Apex of the Pyramid'.[1] He described how senior officials of the then Food, Drink and Tobacco Training Board saw food education in the UK constructed with comparatively small numbers of university graduates, then successively larger layers of lesser skilled people, with the whole structure supported by the so called 'unskilled' workers, who in fact demonstrated varying degrees of skill and needed appropriate training.

A pyramid is, perhaps, no longer the best model to describe the structure of the food industry workforce. The industry has developed from a labour-intensive to a capital-intensive structure; only the relatively small special product sector is less automated. The numbers of operative, craft and supervisory personnel have contracted significantly while degree level employees have maintained or increased their numbers. The pyramid is changing to a truncated cone!

Those responsible for food education have always attempted to keep closely in touch with the changing needs of employees involved with food processing. The food chain has been well served. Nevertheless, significant change in the nature of an industry has clear implications for education complementary to it. The debate about how successfully existing provision meets the needs of student, employee, industry and the public and whether opportunities and challenges to promote professional practice are being taken, is a continuing one.

INITIAL EDUCATION AND TRAINING

Historical Perspective

In the 1940s the need for craft skills and knowledge unique to a specific food commodity was well established in such areas as brewing, baking and cheese-making, and the science and technology of such communities had already attracted much attention from chemists, microbiologists and engineers. Indeed, the first degree course in dairying in the United Kingdom had been developed by the then Reading University College in 1926.[2] Those involved in food education recognised the commonality of food handling operations across different commodity groups and the value of transferable skills and knowledge.

Broad food science and technology courses at advanced level were first developed in the United Kingdom from the late 1940s to the early 1960s. Some grew from commodity-focused courses in breadmaking and dairying, in response both to the shortage of food scientists during the Second World War and to the shortage of students enrolling on some single commodity courses.[2,3] The first degree course in food science was offered by the then Royal Technical College, Glasgow, now the University of Strathclyde, in 1949. The National College of Food Technology was established in 1951 and began degree courses a little later. By the 1960s degree courses had been established at six universities, one polytechnic and the National College. Higher National Diploma (HND) and Ordinary National Diplomas in food science and technology had been developed from earlier commodity-specific equivalents. At about the same time a series of one year post-graduate conversion courses for students from other science disciplines emerged.

Centres such as Hollings College, Manchester, and the College of Food and Domestic Arts in Birmingham continued to provide craft level food commodity courses, mainly in bakery, meat handling and dairying for part-time day release students. These courses were validated by bodies such as the Institute of Meat and the City and Guilds of London Institute.

Most of the provision at advanced level was in food science rather than food technology. A strong secondary education in the sciences was required prior to entry. The thrust of the courses was towards understanding and controlling post-harvest and processing changes in foods. Nevertheless, a clear industry focus was present and students were encouraged, and in some cases required, to gain some first hand experience in the industry.

The demand for research, development and technical or general management personnel has been maintained over the past 20 years, but changes in manufacturing practice reduced the demand for trained craft and supervisory staff. Fewer but larger processing units incorporating automated production and process-control lowered manpower requirements. The amount of educational provision below degree and HND level contracted significantly. This is in contrast to, for example, the enormous growth in courses serving the hotel and catering industry which remains labour-intensive. Only the retail-based baking and meat industries, characterised by their large number of small operating units, have continued to demand a substantial amount of non-advanced education and training.

Current Provision

Tables 1, 2 and 3 show the current provision of food science and technology courses in the United Kingdom. In terms of student enrolment the pyramid referred to by Professor Hawthorn is inverted. Advanced level students constitute 85% of the total enrolment. Degree students are more numerous than those studying Higher National Diplomas. National Diploma student numbers are low. College-based craft-level provision other than for the meat and bakery trades has virtually disappeared. The university sector is the major provider of advanced courses and over a third of United Kingdom universities offer post-graduate research degree opportunities in well defined food science or technology research teams. Recently the number of public sector polytechnics and colleges offering degrees has grown and others are exploring the demand for undergraduate courses.

Combined honours degrees linking food science with subjects such as chemistry, microbiology, nutrition, economics, marketing or hotel and catering have developed. Food marketing is also served by courses which examine all aspects of the food production, processing, distribution and sales chain. Similarly, courses giving greater attention to the management of food manufacture have been introduced.

Obtaining accurate forecasts of the demand for trained personnel is as difficult in the area of food science and technology as it is in any other case. In 1984 the Institute of Food Science and Technology (IFST) attempted a major survey of 2 500 members, many employed in the food industry, in order to build a picture of employment needs. A response rate of under 2% was not encouraging. A further survey by the IFST of education and training needs has recently been completed and is due for publication. More successful was an analysis by the Food Manufacturer's Council for Industrial Training undertaken in 1986 which indicated that between 1986 and 1988 there would be a 13% increase in the demand for qualified food scientists and technologists.[4]

In broad terms the overall supply of potential employees from initial advanced level and post-graduate courses would appear to be in balance or slightly below current industry demand. The employment prospects of students are good. Interestingly, this applies as much to the few full-time craft level students still taking courses in centres such as the Cheshire College of Agriculture as it does for degree level students. A survey of vocational qualifications in the food industry commissioned in 1987 by the Food and Drink Consultative Group of the National Council for Vocational Qualifications is designed to throw light on the demand for qualifications at all levels within the industry. It should be particularly helpful in commenting on operative, craft and supervisory level education and training where, for example, there is some suggestion that the needs of smaller, more labour intensive, processing companies such as diversifying farm-based enterprises, are not being met.

Current Issues

There is considerable variation in the format of food science degree courses but the typical single subject food science degree remains firmly and appropriately built on a science foundation. Students are required to have at least two and often three science GCE Advanced Level passes on entry to a course. Biochemistry, food chemistry, nutrition, micro-biology and food microbiology, physics and biophysics underpin the course, complemented by the study of statistics and information technology. These subjects, in turn, allow food commodity groups, unit operations, process and production control, quality assurance, post-harvest and post-process changes and food legislation to be considered. Superimposed on this content coverage is the use of teaching and learning strategies which aim to develop the generic skills expected of an advanced level science student. Most food science courses require students to augment what they learn in the lecture room and laboratory with work-based experience. In some cases a year long placement is involved, in others eight to ten weeks' employment must be obtained during vacation time. In one course industrial placement is not a compulsory element.

Food technology and food manufacture courses place greater emphasis on business, marketing, management and production aspects of food processing but still devote the majority of their time to a science-based curriculum. Some consider food engineering in greater depth. One year industrial placements are included in all these advanced level courses.

Those providing education frequently question whether the twin beneficiaries of their efforts, the students and professional practice, are adequately served. Where should the balance between meeting their not

TABLE 1
Profile of Education Courses in Food Science and Technology Offered in the UK, 1987-8

Course level	Sector offering course	Mode of attendance[a]	Duration	Number of institutions offering courses	Student enrolments 1986-7	Industrial placement included
PhD	University[b] Public[b]	FT or PT FT or PT	3 yr plus 3 yr plus	15 7	All Higher Degrees: University 182 FT 35 PT Public 32 FT 16 PT	
MSc	University[b] Public[b]	FT PT	1–2 yr 2–3 yr	9 1		
MFC	The Mastership of Food Control is examined by the Institute of Food Science and Technology (UK). No course leads to the award but it is supported by short courses offered by educational and research institutions					
Post-graduate Diploma	University[b] Public[b]	FT FT	1–2 yr 1 yr	4 2		

Course	Institution	Mode	Duration	Number	Enrolment	Sandwich
BSc	University	FT Sandwich	3 yr 4 yr	8 3	566 FT 8 PT	Yes
	Public[b]	FT Sandwich	3 yr 4 yr	1 5	223	Yes
Higher National Diploma[c]	Public	FT Sandwich PT (HNC)	2 yr 3 yr 2 yr	9 2[d]	303 FT[d] 40 PT[de]	Yes Yes
National Diploma[c]	Public	FT PT (NC)	2 yr 2 yr	12 1[d]	220 FT 7 FT[d]	Yes
Craft[f]	Public	PT FT	2 yr 1 yr	2[d]	30[d]	Yes

[a]FT = Full-time, PT = Part-time.
[b]Courses validated by the Council for National Academic Awards.
[c]Courses validated by the Business and Technician Education Council or Scot. Vec.
[d]Figures for England only.
[e]Approximate number.
[f]Only the following are included: CGLI 123 Flour Milling.
CGLI 125 Dairying.
CGLI 129 Food and Drink Industries Certificate in Processing and Quality Control.

TABLE 2
Range of Food Science and Technology Courses Offered in the UK, 1987–8

Course level	Course title	Universities		Public sector education	
		No. of courses	*No. of institutions*	*No. of courses*	*No. of institutions*
PhD	Wide range of research topics undertaken		15		6
MSc	Food Science	4	9		1
	Food Technology	1			
	Food Engineering	2			
	Food Analysis	—		1	
	Meat Science	2			
	Brewing	2			
	Agriculture + Food Marketing	1			
	Agriculture + Food Biotechnology	1			
Post-graduate Diploma	Food Science	1	4		2
	Food Technology	—		2	
	Food Composition + Processing	1			
	Food Resources	1			
	Brewing	1			

BSc	Food Science	6	1	10	6
	Food Technology	2	1		
	Food Manufacture	—	1		
	Food + Biochemistry	1			
	Food + Marketing	1	1		
	Food + Management	1			
	Food + Mechanical or Chemical Engineering	1			
	Food + Microbiology	1			
	Food + Physiology/Nutrition	2	2		
	Agriculture + Food Marketing	2			
	Agriculture + Food Microbiology	1			
Higher National Diploma	HND Science (Technology of Food)		9		9
National Diploma	ND Science (Technology of Food)		12		12
Craft (England only)	Flour Milling		1		1
	Dairying		1		1
	Food and Drink Industries Certificate in Processing and QC		1		1

A. J. Youngs

TABLE 3
Further and Higher Education Institutions Offering Courses in Food Science and Technology in the UK, 1987–8

Institution	PhD	MSc	Post-graduate diploma	BSc	HND	ND	Craft
Universities							
Bath	X				(Food Microbiology)		
Belfast	X			X			
Birmingham	X	X			(Brewing)		
Bristol	X	X			(Meat Science)		
East Anglia	X						
Herriot Watt	X	X	X		(Brewing)		
Leeds	X	X	X	X			
London (Kings)	X	X	X	X			
Nottingham	X	X		X			
Newcastle	X			X	(Agriculture and Food Marketing)		
Reading	X	X		X			
Strathclyde	X	X	X	X			
Surrey	X			X			
Ulster	X			X			
Wales (Aberystwyth)	X	X		X	(Agriculture and Food Marketing)		

Institution		(Food Marketing)	(Catering and Applied Nutrition)
Public sector higher education			
Huddersfield Poly.	X		
Humberside CHE	X		X
Manchester Poly.	X		X
Oxford Poly.	X		
Sheffield City Poly.	X		
South Bank Poly.	X	X	X
Seale Hayne College	X	X	X
South Glamorgan IHE		X	X
Public sector further and higher education			
Birmingham College of Food		X	X
Glasgow College of Food		X	X
Loughry College of Agriculture		X	X
West of Scotland College of Agriculture		X	X
Reading College of Technology		X (HNC)	
Public sector further education			
Cheshire College of Agriculture			X
Lowestoft CFE		X	X
Nottingham College of Agriculture		X	X
Somerset College of Agriculture		X	X
Thomas Danby College		X	

always convergent needs be struck? Do courses equip students with the competences required in employment? What is the role of the employer in developing the recent graduate in a first post?

One full exploration of higher education in food science and technology in Europe conducted in 1979[5] considered these points in detail, but several topical issues continue to keep the curriculum debate on the boil. Pressures exist to bring additional elements into the curriculum and to introduce new emphases in order to further strengthen the relevance of courses. At the same time student backgrounds prior to joining courses are changing and may influence the starting point of advanced courses.

Education consistently gives a high priority to the long term needs of students, as individuals facing decades of change in professional practice. Others, including industry, may strike the balance between short and long term perspectives differently. The general drift of the debate emphasises the extended nature of education as a process which must continue beyond initial full-time courses, and the need for the individual, the education service and employment to work as a partnership using the best market knowledge available.

Curriculum content

The need for a strong science foundation for food science and technology cannot be compromised. In recent times nutrition and information technology have been added to the foundation; in the case of the former not without some resistance. Consumer-led pressure, relating to food preservation, processing and quality, is one of the latest issues which stimulates curriculum review, the outcome of which is often to suggest an extension of subject coverage. An essential concern of curriculum review is to determine whether broad, multi-disciplinary, multi-commodity courses are rigorous enough in each area covered and, at the same time, not so wide ranging that students are unable to cope.

Industrial experience and student competence

A range of views exists about the value of industrial experience in helping students to develop understanding, competence and ability to make an early contribution on entering employment. The question of post-course competence has been intensified in the UK by the review of vocational qualifications commissioned by government and the subsequent establishment of the National Council for Vocational

Qualifications. The Council is charged to encourage the development of employment-led vocational qualifications which guarantee not only the acquisition of skill and knowledge but also the ability to apply that skill and knowledge in an industrial setting and to standards set by industry.

A sandwich year of industrial experience is included in all but one non-university degree course and is optional in Higher National Diploma courses. Sandwich experience presents an extra dimension allowing competence to be developed, demonstrated and assessed. The Training Commission's funding of industrial attachments of between 8 and 24 weeks administered by the Food Manufacturer's Council for Industrial Training encourages the industry to provide such experience. The full potential of industrial placement as a planned and integral part of the course is not, however, always realised. To achieve this potential it is necessary for the host company or organisation to be actively involved in developing the student's potential and in assessing performance.

Commercial awareness

The Higher Education Enterprise Initiative of the Training Commission will provide an average of £1M for each university and polytechnic over the next five years to encourage higher education institutions to ensure that all students at undergraduate level receive some exposure to 'enterprise' in the curriculum. Food technology courses are perhaps further down this road than food science courses but on both one way a 'feel for business' may be developed is through an enhancement of existing subject coverage. Assigning a student a company and asking that the fortunes of the company be followed through the financial and trade press so that a review can be presented at intervals is, for example, one basic technique which provides insight into the commercial environment while reinforcing wider professional knowledge of the industry.

Student background

Those providing HND and degree courses have always had something of a battle to reach school pupils in order to tell them about food industry courses and careers. As a result of demographic change it is estimated that the number of 18-year-olds likely to seek entry to higher education between 1985 and 1995 will fall by between 14 and 23%.[6] If a constant or increased number of students are to be enroled on courses a greater spread of academic ability, age and previous education background will be necessary. This applies particularly to achievements in the sciences. Will, for example, pathways be needed to accommodate greater numbers of mature students? Should students whose primary qualification at GCE Advanced Level is home economics be considered for food science courses?

Added to this change in student numbers is an evolution in pre- and post-16 education which amongst other things is encouraging less subject specialisation. In some subjects, national diploma qualifications are gaining currency as an alternative to GCE Advanced Levels. The impact of the national foundation curriculum on breadth and balance within the 5–16 curriculum, of introducing Advanced Supplementary Level courses to allow students to study a broader range of subjects up to the age of 18, and of changes in the methods of teaching and assessment in GCSE courses all require that the curriculum of advanced courses be examined and if necessary modified to fit the different experiences of students.

Industry liaison

Dialogue has long existed between education and the food industry through such forums as college departmental advisory committees or boards, the research and consultancy activities of lecturers, student industrial placements and representative trade bodies such as the Food Manufacturer's Council for Industrial Training or the Biscuit, Cake, Chocolate and Confectionery Alliance. Professional bodies such as the Institute of Food Science and Technology, the Institute of Food Technologists and the Society of Chemical Industry have also fostered contact between professionals and helped education in its role of taking the initiative in interpreting the needs of industry and employees. Education always complains of difficulty in obtaining a consensus view from industry. Greater efforts could, however, be made by course administrators to obtain constructive comment from past students in their early years of employment in order to inform reviews of course content and delivery.

Each of the issues identified above suggest that more than ever, education needs a clear view of the needs of its clients, both students and employers, and of how the needs should be prioritised. Within hotel and catering education, standing conferences of college heads of department successfully act as co-ordinators of education's view in discussion with other bodies. Perhaps it is appropriate to ask whether a case now exists for bringing together all providers of education in food science and technology to undertake a similar role.

POST-EXPERIENCE EDUCATION

The role of post-experience education is likely to increase in relevance as pressures mount on the breadth and perhaps the length of initial courses. The education service, research associations, individual companies, professional bodies and the nine non-statutory training organisations serving the food and drink industry all make provision for post-experience education in technical subjects for the food industry. Most of it is of short duration and sector- or company-specific, and so, difficult to map. How well this provision meets need is similarly difficult to assess.

It is helpful to differentiate between certificated courses available to the individual who wants, often as a result of self-motivation, to improve qualifications, and that provision made to meet specific company or sector goals and usually not certificated.

The education sector now makes little part-time advanced level food science or technology course provision relevant to the large scale food processing industry. The few part-time Higher National Certificate courses and the part-time MSc course in food analysis at the Polytechnic of the South Bank are notable exceptions. The only other major post-experience qualification is the Mastership of Food Control examined by the Institute of Food Science and Technology with the support of the Royal Society of Chemistry and the Institute of Biology. Students are required individually to prepare themselves for the examination and finally to prepare a thesis on an industry-related topic. Nevertheless, they are encouraged to attend short courses offered by a variety of bodies. The low number of students following the programme, together with the absence of other part-time provision, would appear to reflect the industry as one offering little progression to less qualified employees. There is a growing gap between operative level and highly trained technical employees. Few people are climbing the qualifications ladder.

Nationally there is significant current interest in promoting wider access to further and higher education for mature students. Those adults willing or able to follow a full-time course have available the full range of initial education courses and it is not uncommon to find one or two mature students on such programmes. For the remainder who might seek qualifications through open access or distance learning schemes there are no advanced level food-specific courses available other than the Food Production Systems module of the Open University.

Technical education and training to meet specific company needs, particularly at operative level, is undertaken primarily by companies and non-statutory training organisations. In addition, colleges contribute as the contracted provider of courses. The Cheshire College of Agriculture, for example, maintains two full-time members of staff specifically to undertake short courses for the food industry. The bakery school of the Polytechnic of the South Bank has a sound record of providing company- and sector-specific sponsored courses.

For professional staff important supplementary provision is made by the professional bodies, research organisations and education through conferences, symposia and short courses.

The research associations have a good range of courses and other independent short courses such as the ice cream course at the University of Reading and the food microbiology course at the University of Surrey are long established. Food engineering is one specialist area where little short duration post-experience education is available but where several organisations are hoping to offer courses.

In spite of the good record created in many cases, not all short courses offered by education are judged to be relevant to need. A recent study of continuing professional development in the process and process plant industries[7] concluded that although external short courses offered by higher education remain a staple diet of professional development they are perceived by companies as being insufficiently practical and too remote from the real issues. Relevance is a specific company need; individual participants would add the need for accreditation towards a qualification.

One of the aims of the Department of Education and Science when it established the Professional, Industrial and Commercial Updating Initiative (PICKUP) was to identify more clearly and then meet post-experience education and training needs. One vehicle used has been the establishment of short-life Local Collaborative Projects in which employers and education, usually public-sector, set out to identify updating needs and produce a long term strategy to meet them. One of the most successful PICKUP projects has been the North-West Food

Technology Consortium, now permanently established at Salford College of Technology. Four north-west colleges and 18 north-west food and drink processing companies organised a range of courses to meet the training needs of personnel in all categories of activity. This process is now continuing under the management of a permanent secretariat and a similar project has been initiated in the East Midlands and one undertaken in the London region. An excellent model for meeting needs through an education and industry partnership has been established.

A major challenge exists within post-experience food education and training to develop modes of delivery which fit working patterns, meet commercial and industrial constraints and facilitate participation rather than presenting hurdles. Open learning approaches are being increasingly used within industry to deliver training in a flexible way, at times convenient to employer and employee and in a way which is not so dependent on a minimum student group size. An emerging lesson is, however, that open learning needs a high level of tutorial support, particularly for those students who are engaged on a long term programme. The retail industry including food retail is one sector which has led the way in the use of distance learning to meet limited objectives. The education sector has been an important contributor to course material in some cases. So far, however, the food industry and education have not engaged in this form of provision to any great extent although some Open Tec programmes on hygiene, legislation and chocolate manufacture do exist.

CONCLUSIONS

Educational provision oriented to the needs of the food processing industry has changed significantly since the 1940s. The emergence of advanced level courses at that time reflected the growing technological sophistication of the industry and the demand for informed management, research and development support. Technical advance promoted automation and centralisation and the change from a labour-intensive to a capital-intensive industry. Moulded by these changes, the education service in the late 1980s is called upon to make only limited provision at craft and supervisory level. There has however been an expansion of advanced level courses offered by public sector institutions to join the well established university provision.

Although clear messages about the level and nature of demand for personnel entering the food industry are difficult to obtain, the overall supply of potential employees from HND, degree and post-graduate courses would appear to be in broad balance with industry demand. New challenges exist, however, for education to continue some supervisory, craft and operative level education and to promote post-experience education.

The role of continuing education may increase in relevance as new pressures affecting initial education fuel the debate about course breadth and length. Changes in student backgrounds, the increasingly wide profile of expertise demanded of the professional food scientist and technologist and the related issues of competence as an end product of education are all likely to emphasise education as a continuing process and not one which ceases with the beginning of employment.

Fundamental to the responsiveness of education is the way it is marketed. The needs of both student and professional practice must be met, but education must take the initiative in detecting and meeting their needs. Scope exists for constructive comment to be obtained from past students in order to inform course content and delivery. Of continued importance is the nurturing of strong college–employer networks.

Emil Mrak concluded in his contribution to the symposium on 'Education in Food Science and Technology' in 1970[8] that this field is a diverse and rapidly changing one and because of that an exciting one in which to be involved. It continues to be so.

REFERENCES

1. Hawthorn, J. (1970) Proc. Inst. Fd Sci. and Technol. (UK), 3, 134.
2. Rothwell, J. (1979) J. Soc. Dairy Technol., 32, 105–8.
3. Hawthorn, J. (1979) In: Management Training in Food Industries — Higher Education in Food Science and Technology in Europe, J. Lenges (Ed.) European Federation of Chemical Engineering, Brussels, III-A-1.2.
4. Anon. (1986) FMCIT Annual Report 1986, The Food Manufacturer's Council for Industrial Training, London.
5. Lenges, J. (Ed.) (1979) Management Training in Food Industries — Higher Education in Food Science and Technology in Europe, European Federation of Chemical Engineering, Brussels.
6. Anon. (1986) Projections of Demand for Higher Education in Great Britain 1986–2000, Department of Education and Science, London.

7. Geldhart, D. and Brown, A. S. (1987) *A Largely Satisfied Need: Continuing Professional Development for Process and Process Plant Industries,* Further Education Unit, Department of Education and Science, London.
8. Mrak, E. (1970) *Proc. Inst. Fd Sci. and Technol. (UK),* **3**, 130.

Educational Transfer in Food Technology

GEOFFREY CAMPBELL-PLATT

Department of Food Science and Technology, University of Reading, UK

ABSTRACT

The world population has reached 5 billion people, an increasing number of whom live in cities, away from the means of producing their own food. These urban citizens expect to be able to purchase their food requirements from retail stores conveniently situated to their homes or workplaces. Basic food needs must be satisfied, but increasing prosperity brings with it the demand for a wider range of quality food products. Wide variety and year-round availability of foods which are fresh or subjected only to minimal methods of preservation will be expected increasingly. It will also be assumed that all foods offered in food service are safe to consume, and those offered for sale are safe and wholesome, even after periods of storage in consumers' homes.

These demands for a wide range of minimally-processed, safe, convenient food products attractively presented can only be met by the involvement of professional food scientists and technologists. Awareness of this subject area is presently low in schools, and a new initiative to increase interest has been taken in the establishment of an introductory course in food technology for sixth formers, at the University of Reading.

A high proportion of qualified food technologists enter the food industry, including the increasingly important area of food distribution and retailing. There, they play a major role in research and development in a fast-moving industry, which responds rapidly to changing consumer preferences. They help deliver good-quality food products safely to consumers through implementing good manufacturing practice, hazard analysis and quality assurance. Their personal and professional roles in educational transfer in food technology will be vital in maintaining a healthy well-fed growing world population in the next century.

113

INTRODUCTION

Food technologists are concerned with the production of sufficient quantities of good quality, attractively-presented food products, and are responsible for ensuring that these products conform with advisory and legal requirements, and are safe to eat.

With a world population of 5 billion, an increasing number of whom live in large cities, the production of a regular safe food supply cannot be left to individual effort, but requires a major educational and training programme in food science and technology. The professional food technologists will need to have a broadly-based education in the sciences of chemistry, physics, microbiology, engineering, and processing technology as applied to food. Further, they will need to know about management, be commercially aware, and to demonstrate the personal leadership and communication skills needed to apply their knowledge successfully.[1]

CONSUMER AWARENESS

All members of the public from children to the aged are food consumers. Children's eating habits are conditioned by parents, homes, culture, schools, and the influence of their whole environment, including advertising, choices available and peer-group pressures. As an important part of their education, children need to be made aware of the importance of eating the correct quantity of a well-balanced diet.

In addition to learning about the basic food components, and their significance in supplying energy and vital nutrition to growing children and adults, there should be awareness of particular problems that arise in different parts of the world. For example, in much of Africa and parts of Asia and Latin America, where food shortages occur, children need to be made aware of how to ensure sufficient energy intake. In more affluent western countries, with their increasingly sedentary lifestyles, food choice needs to be more selective to prevent overnutrition.

Good habits in food selection, food preparation, personal and food hygiene, and food storage and consumption need to be established in childhood. The total food chain from agriculture through food product manufacture, distribution, retailing and food service is the largest employer of people in most economies. Everyone is a food consumer, at the individual level. The future success of a safe dependable food chain internationally will need not only highly-trained food technologists but also a literate, educated and aware population.

In western countries, increased prosperity and life expectancy, coupled with abundant reasonably-priced food supplies, has led recently to closer examination of the effect of diet on health.

In Britain, for example, with its high incidence of cardiovascular disease, health and nutrition advisers, with expert committees, have addressed the subject and have produced a series of educational guidelines. These include *A discussion paper on proposals for nutritional guidelines for health education in Britain* (NACNE, 1983),[2] *Diet and cardiovascular disease* (COMA, 1984),[3] *Eating for a healthier heart* (JACNE, 1985),[4] and *Guide to healthy eating* (Health Education Council, 1986).[5] The last of these publications, aimed at the whole population, states that the best chance of living a fit and healthy life is by eating a good balance of food, taking regular exercise and not smoking. The guide concludes: 'Cut down on fat, sugar and salt, eat more fibre-rich foods, eat plenty of fresh fruit and vegetables, go easy on alcohol, and get plenty of variety in what you eat'. Alongside these public-awareness reports, some sectors of the agricultural and food industries are trying to increase awareness of food to schoolchildren, for example by the publication of educational material and *Dairy Education* for schools by the National Dairy Council in Britain.

INTRODUCTORY COURSE IN FOOD TECHNOLOGY FOR SIXTH FORMERS AT THE UNIVERSITY OF READING

In a move to meet future needs, and to increase the profile of the subject in Britain's schools, at the University of Reading we initiated in 1987 a week-long introductory course in food technology for lower-sixth formers. All 52 places on the course were fully sponsored by the food industry, who publicised the offer of a place on the course through their local schools throughout the country. Some food companies set competitions, most interviewed students interested in attending the course, and several offered periods in their food companies in conjunction with attending the course at Reading. Participants were required to give a report on the course to their fellow sixth formers and next year's lower-sixth formers back in their own schools, in the presence of, and with help as required, from their sponsors. Publicity in association with the course was obtained with local news media, through company newsletters, and subsequently, through the *'Annual Report of the Food Manufacturer's Council on Industrial Training (FMCIT)',* who strongly welcomed the initiative.

The first course was run in July 1987, towards the end of the school summer term, and after the end of the university term, allowing the students to all be accommodated in a university hall of residence on campus. Three third or final year food technology undergraduate students from the university lived with the sixth formers to participate in student campus life, as well as attend the course.

During the programme (Table 1) as much contact as possible was established with recent food technology graduates, allowing the sixth formers to receive a first-hand view of the course and careers opportunities from other young people. This aspect was particularly highly appreciated by course participants, as revealed in the questionnaires completed at the end of the course. The participatory events, debates, practical classes and the production game were all identified as highlights, and will be extended and expanded in future courses, with team projects introduced to run throughout the week.

The course demonstrated to the participants the opportunities in industry generally, and the food industry in particular. In several cases, it was the young persons' first time inside a factory. One written comment afterwards was illuminating:

> Personally, I hadn't seriously considered going into industry before, particularly the food industry, but now I feel confident that I will probably apply for this course, as it certainly opened my eyes, and the main thing, it fired my enthusiasm towards this course in particular.

The questionnaires revealed that among the participants, some two-thirds had previously considered a career in food, but only a quarter had considered undertaking a university course in food technology. At the end of this introductory course, which all found helpful and all but one (who found it interesting in parts) found interesting, 45% said they would be applying to the university to take the course in food technology, a further 45% were uncertain, and the remaining 10% were not qualified to do so. Several of this latter group were studying combinations of sciences and arts to General Certificate of Education Advanced Level. The new course in BSc Food Manufacture, Management and Marketing which is planned to start at Reading in 1989 would allow most of these candidates to study for a degree in food in the future,

TABLE 1
University of Reading Introductory Course in Food Technology for Sixth Formers

Day 1	
a.m.	Arrival, lunch and welcome
p.m.	Lecture: Food technology and the food industry
	Tour of Food Studies Building (student guides)
Evening	Tour of university campus (with students)
	Food hygiene film, with discussion
	Debate by sixth formers on current food product issues (particularly additives)
Day 2	
a.m.	Lecture: Introduction to food chemistry
	Practicals: Group A — food chemistry laboratory
	Group B — food processing laboratory
p.m.	Lecture: Introduction to food microbiology
	Practicals: Group A — food microbiology laboratory
	Group B — food chemistry laboratory
	Visit to food retail store, with buffet, and talks from recent Reading graduate food technologists
Evening	Food production game by teams of sixth formers
Day 3	
a.m.	Lecture: Introduction to food processing
	Practicals: Group A — food processing laboratory
	Group B — food microbiology laboratory
p.m.	Visits to food factories, with discussion by food technologists working in factories
Evening	Formal course dinner, with talks by University Vice-Chancellor and guest speaker looking at future of food industry
Day 4	
a.m.	Careers in the food industry — talks by recent graduates of food technology courses at the University of Reading
	Closing questions, answers and discussion
	Lunch, photographs and departure

if they wished. The sixth formers, university students and staff, and the industrialists were unanimous in having found the introductory course a worthwhile and enjoyable experience. The industrial sponsors and staff found the company and questioning of the sixth formers to be challenging and stimulating. It has been agreed to run the course in future on a regular annual basis. The effects on increasing numbers of good quality applicants will need to be assessed over a longer timespan, but already initial results are encouraging. From this first introductory course, seven have subsequently joined the University of Reading course in Food Technology.

THE TRAINING AND ROLE OF PROFESSIONAL FOOD TECHNOLOGISTS

The study of food science and technology as a degree subject in universities is a relatively new discipline. Early courses started in the United States of America[6] and in Britain[7] in the middle part of this century. Perhaps because it is not taught as a school subject, and its relatively recent appearance at universities and colleges of higher education, it has not been a popular choice of school leavers. Moore[6] noted that high-school educators in the United States of America had a view of food science slanted towards the home economics areas and were not well aware of food processing and technology opportunities. In the author's own experience in Ghana, West Africa, most students found their way into the subject after first applying unsuccessfully to study medicine, while in Britain awareness among university applicants has long been low. The food industry has demonstrated a strong continuing demand for food scientists and particularly the food technologists with their experience of industry gained through sandwich degrees or industrial training placements.

While education and basic science is universal, the successful application of the knowledge through technology will depend on an understanding of technology transfer in a particular local environment. Failure to understand and acknowledge local conditions and needs can lead to much frustration and waste of effort. It is to achieve this close harmony that industrial training during degree courses, such as in the food technology and biotechnology degrees at the University of Reading, is particularly important. Periods in industry allow students to explore the ways and means of achieving results in industry, and open the eyes of manufacturers to new ideas and put educational transfer to practical test in industrial development. Food development is an international issue, it knows no boundaries, and every nation is interdependent.

As illustrated in Table 2, many students from developing countries continue their education in food science and technology, often through PhD training in developed countries. To help this educational transfer

TABLE 2
Employment of Food Science and Technology Graduates — %

Source: Graduates:	Food technology departmental records University of Reading	Anon. (1980)[9] USA	Campbell-Platt (1980)[10] Ghana
Research and development	32	25	29
Quality assurance and control	10	18	21
Production management	29	5	5
Other — marketing, information, administration	4	27	5
INDUSTRY TOTAL	75	75	60
Continuing education	6	10	29

go successfully for mutual benefit, the International Union of Food Science and Technology (IUFoST) has drawn up guidelines for PhD training for candidates from developing countries.[8] These point out that the research topic selected should be as directly relevant as possible to the candidate's country's needs or potential, and the research programme should be structured so that research can be continued with facilities and resources available in the developing country. This can be achieved by the candidate spending a portion of time in the middle of the research programme in his or her own country, preferably with a supervisory visit during this time. This should ensure that the supervisor understands the local significance and need for the research, and the candidate is able to introduce the work successfully into the country's development.

As consumer demands for 'fresher', less processed food increases, the world demand for foods processed by minimal methods of preservation will continue to increase. For safe development and production of these foods there will be an increasing need to apply combinations of milder technologies, involving improved hygiene, control of processing and distribution through to final consumption. In most developed richer countries the climate is generally cool, and distribution systems are developed and efficient. In the early part of the next century, as the demand for these products increases steadily in the hotter developing countries with less infrastructure, it will require the combined resources of food technologists working together internationally to deliver safely a wider range of convenient minimally-processed foods to consumers worldwide.

The knowledge which food science and technology graduates have acquired is transferred through their employment. Compared with most other university subjects, a high proportion of food science and technology graduates enter industry, including retailing. For example, for the University of Reading, recent overall figures for graduates show 14% entering industry, plus a further 23% entering commerce, including retailing.

As seen in Table 2, the percentage of food technology graduates from Reading entering industry, over the twelve-year period 1974–86, was 75%. Figures for comparison from the United States of America[9] show a similar percentage entering industry, while figures for the twelve-year period 1966–78 in a developing country, Ghana, show 60% of graduates entering industry.[10] The lower percentage figures for Ghana are counterbalanced by much higher numbers continuing their education, either short- or long-term, with many students undertaking further studies in developed countries. Ideally, the experience gained would be translated back to the home country at a later date.

While such high percentages of graduates do enter the food industries, the degree is broadly based and allows graduates to branch off into many other fields. These may include working in government or many other careers — including recently from Reading, accountancy and the police. Some of the women graduates leave the employment area for a period while bringing up children at home. Home and personal contacts are another important area of educational transfer. A survey of Ghanaian students revealed that through studying food, over 90% had made changes in their own practices during the course, and all had suggested changes in practices to friends and relatives. Half of the students had suggested changes to those involved in food production or service so helping to increase the transfer of education. Most of the students reported that only some of their suggestions had been accepted by others, but few found that no changes were accepted. This individual effort is an important informal part of the education transfer chain, and as more people study food it will help in the important area of increasing 'food literacy' in populations in all countries, developing and developed.

Most cases of microbiological food poisoning, the major safety hazard with food products, are due to mishandling by people. The majority of these cases arise through bad practices in food service establishments and homes[11] (Table 3), which are the most difficult areas to put under the direct control of food technologists. The importance of

TABLE 3
Microbiological Food Poisoning
Origin of Outbreaks in Canada 1973–7 (Todd, 1983)[11]

Identified source of problem	% of cases
Food service (restaurant, etc.)	55
Homes	12
Food processing plant	3
Retail outlets	2
Raw material source (farms and dairies)	1
Distribution	0·2
Not identified	27

greater personal awareness, understanding at the individual level cannot therefore be overstressed.

RETAILERS, GOOD MANUFACTURING PRACTICE AND FOOD CONTROL

In recent years in Britain there has been a major increase in the dominance and importance of major food retailers, with the majority of food product sales passing through the stores of just six major food retailing groups. These retailers need the services of well-qualified and able food technologists. The number of technically-qualified staff employed by these companies has increased from 164 in 1970 to 391 in 1985.[12] The majority of these staff are graduates; indeed the largest employer of University of Reading's food technologists over the years has been Marks and Spencer plc. The retailers, like manufacturers also employ people with diplomas and other technical qualifications. Marks and Spencer have led the way in setting high standards for product specification, hygiene and quality and have shown that in a more prosperous society where there is an ample food supply, value through quality is increasingly more important than just price in determining consumer purchases. Other retailers have followed this lead, and have through their own-label products shown their ability to respond rapidly to changing tastes and to innovate in product development. Retailers have invested heavily in efficient central-distribution systems which allow consumer requirements for fresher short-life minimally-processed food products to be met. The retailers' strong technologically-based presence continues to increase in importance in Britain, and this pattern is being followed in other countries. Food service operations are also coming more under central, international control, often lead by American fast-food chains. Successful modern food companies and retailers need to be able to develop and produce reliably and safely a wide range of new food products. To enable this to happen these companies employ graduate food technologists who can transfer their education in food technology to the food producers. These graduates or diplomatists find themselves responsible for safety, hygiene, procurement of raw materials, product specifications, technical innovation and product development, and conformation to labelling and compositional legislation. Education is an ongoing process, and the food scientists and technologists will be expected to further their knowledge and experience through active membership of one or more professional societies. Particularly relevant professional societies include the Institute of Food Technologists (IFT) in the USA, and the Institute of Food Science and Technology (IFST) and the Society of Chemical Industry (SCI) Food Group in the UK. The practising technologist will need to keep up-to-date by reading current literature, attending symposia, and short courses run by the professional institutes, universities, colleges and research associations or institutes, both to acquire new knowledge and to exchange ideas. Participation in the production of codes of practice, advisory committee reports or professional publications is a particularly good way of contributing and gaining knowledge and sharing expertise. The production of the recent code of practice *Food and Drink Manufacture — Good Manufacturing Practice: a Guide to its Responsible Management*[13] by the IFST is a good example of such an exercise. It is aimed at ensuring that food products are consistently manufactured to a quality appropriate to their intended use, covering both manufacturing and quality control procedures.

In order to be effective in transferring and applying knowledge in food technology, one needs particular personal skills, combined with technical competence. Management should be taught as part of food science and technology courses. The paramount skill is of communication and team work. Self-motivation, interest and ambition should provide the necessary dedication to producing the personal and collective commitment towards product quality and safety. This example needs to be demonstrated by the organisations' leaders. When in 1971 the pioneering Pillsbury Company introduced the hazard analysis critical control point (HACCP) system for its food processing operation, the managing director decided that nobody on the staff would gain promotion or a salary rise unless they could positively demonstrate their commitment to safety and quality.[14] Leadership is essential and the successful companies will be those which practise good management.[15]

CONCLUSION

The successful food technologist needs to be decisive but flexible, innovative, to be commercially aware, and to seek responsibility. Above all, one needs to have a sense of enjoyment and fulfilment in the job, and to receive personal satisfaction from seeing new or established products produced to high standards of quality and safety. If something does go wrong, it is no longer one's personal family or friends who are at risk, it is a much wider range of consumers, nationally and internationally, who take it for granted that what they eat will be safe, as well as nutritious. The dedication to producing safe products needs to be just as great by food technologists as that by aircraft technologists in ensuring that their planes are safe for regular passenger flights. In both cases, it is the professionalism and commitment of dedicated technologists who ensure an enjoyable and safe experience for the consumer. A challenging, important and worthwhile career awaits those who have the commitment to the future wellbeing of the world's growing population.

REFERENCES

1. Bauman, H. E. (1979) Changes in industrial research require changes in graduate education. *Food Technol.,* **33** (12), 30–1.
2. NACNE (1983) *A discussion paper on proposals for nutritional guidelines for health education in Britain,* Health Education Council, London, 40 pp.
3. COMA (1984) *Diet and cardiovascular disease. Report on health and social subjects No. 28,* Dept of Health and Social Security, HMSO, London, 32 pp.
4. JACNE (1985) *Eating for a healthier heart,* British Nutrition Foundation and Health Education Council, London, 27 pp.
5. Health Education Council (1986) *Guide to healthy eating,* Health Education Council, London, 30 pp.
6. Moore, A. B. (1976) Attitudes about food science among high-school educators. *Food Technol.,* **30** (4), 76–82.
7. Aylward, F., Hawthorn, J., Lawrie, R. A., Rolfe, E. J. and Ward, A. G. (1971) Food science and technology in the universities of the UK. *Chem. Ind.,* (37), 1030–5.
8. IUFoST (1980) Guidelines for PhD training for candidates from developing countries. *IUFoST Newsl.,* (2), 1–4.
9. Anon. (1980) The Institute of Food Technologists' 1979 membership salary survey. *Food Technol.,* **34** (1), 28–36.
10. Campbell-Platt, G. (1980) Food science and nutrition training in Ghana — the first hundred graduates. *Proc. Inst. Food Sci. Technol. (UK),* **13**, 63–70.
11. Todd, E. C. D. (1983) Foodborne disease in Canada in a 5 year summary. *J. Food Protection,* **46**, 650–7.
12. Senker, J. M. (1987) Food technology in retailing — the threat to manufacturers. *Chem. Ind.,* 20 July, 483–6.
13. IFST (1987) *Food and drink manufacture. Good manufacturing practice: a guide to its responsible management,* Inst. Food Sci. Technol., London, 60 pp.
14. Bauman, H. E. (1987) Food quality, management and marketing. *Proc. 7th Inter. Congr. Food Sci. Technol.,* Singapore, 27 Sept.–2 Oct. 1987.
15. Campbell-Platt, G. (1987) Practising good management. *Food,* **9** (7), 45–7.

The Educational Needs of the Catering Industry with Special Reference to Social Services Catering

MAUREEN MALIK and ANN WEST

Department of Hotel and Catering Studies, Huddersfield Polytechnic, UK

ABSTRACT

The catering industry in the UK is rapidly growing and currently employs 2·5 million people. It is characterised by having a high labour turnover, low wages and a largely unqualified workforce.

The industry is diverse, ranging from small family run hotels to large central cook–chill production units, operated on factory lines producing thousands of meals per day. Technological change is advancing, moving the emphasis from traditional craft methods to processed foods produced by caterers or food manufacturers and involving increasingly complex technology. This approach is typified by the continually growing fast food sector.

Social services is a neglected part of the catering industry in which little professionalism is employed, yet it provides an essential service in the community feeding vulnerable groups, such as the elderly. In the future, this must alter and new systems need to be developed, operated and managed by a scientifically educated workforce.

Thus catering education must change in two ways. It must attract more people into its courses and provide an education which reflects and predicts the changing complexity of the catering industry.

CATERING EDUCATION — THE OVERALL SITUATION

Traditionally the catering industry is divided into three sectors. The first one is direct profit making, where the consumer pays profit directly to

125

Catering Industry

Direct Profit Making Operations	Indirect Profit Making Operations	Non-profit Making Operations

e.g. Hotels
    Restaurants
    Fast Food Outlets
    Take-aways
	e.g. Workplace Feeding
	    School/College
	    Feeding
	    Hospital Feeding
		e.g. Social Services
		    (Old People's
		    Homes, Meals-on-
		    Wheels, Children's
		    Homes)
		    Hospitals
		    Schools/Colleges
		    Armed Forces

FIG. 1. Categories of the catering industry according to their profit making characteristics.

the operator. Secondly, indirect profit making, where the consumer may contribute some direct payment, but the operator's profit will be paid by a host organisation and thirdly the non-profit making sector where the consumer makes no direct payment and the host organisation operates the facility as a necessary service. Examples from these different categories of the catering industry are given in Fig. 1. However, these days the divisions are becoming increasingly blurred as operations in a non-profit making sector are increasingly expected to be non-loss making.

The catering industry in the UK is growing rapidly and currently employs two and a half million people in a wide range of sectors. This represents 10% of the UK workforce, approximately half of whom are employed in the commercial sector in hotels, restaurants, pubs, clubs and contract catering. The remainder are employed in schoolmeals, hospitals, social services and similar types of non-profit making employment. There is a predicted 2·7% annual increase in employment in the commercial sector to take place between 1985 and 1990. Since the beginning of the decade 30 000 new jobs have been created. This figure takes no account of leavers. In addition, 33 000 managers and supervisors will be needed by 1990. The first year (1987) entry to higher education courses in the subject area are:

Degrees	1033
Postgraduate conversion	91
HCIMA Part B	346

HND	1805
BTEC National Diploma	2921
Total	6196

(Source: HCIMA)[10]

Hence, the mis-match between the industry's requirement and the college output is very large. The industry is characterised by having a high labour turnover and a large number of part-time employees, most of whom are women, and its major growth areas are in the fast food, pub catering and medium priced restaurant sectors.

Within each catering sector there are differences in scale and complexity and the examples in Fig. 2 illustrate some of these variations. Thus the requirements of catering operations along this spectra vary and the catering industry is made up of many different combinations of factors coupled with consumers who make complex demands on caterers. Therefore, in considering the educational requirements of the catering industry many facets have to be taken into account encompassing aspects of the social sciences and the natural sciences. We should note that the growth areas in catering are those which rely increasingly on the use of processed and pre-prepared products from the food manufacturing industry. Alternatively, some caterers are now producing their own pre-prepared products in large central production units, examples of this trend are seen in the number of cook–chill units opening in hospitals.

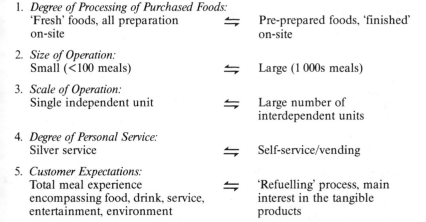

1. *Degree of Processing of Purchased Foods:*
   'Fresh' foods, all preparation on-site ⇌ Pre-prepared foods, 'finished' on-site

2. *Size of Operation:*
   Small (<100 meals) ⇌ Large (1 000s meals)

3. *Scale of Operation:*
   Single independent unit ⇌ Large number of interdependent units

4. *Degree of Personal Service:*
   Silver service ⇌ Self-service/vending

5. *Customer Expectations:*
   Total meal experience encompassing food, drink, service, entertainment, environment ⇌ 'Refuelling' process, main interest in the tangible products

FIG. 2. Variations in scale and complexity of catering operations.

This demands a close relationship between the food manufacturing and catering industries as the degree of processing which the food receives before entering catering affects the basic requirements of the catering system and the type of skill required.

If foods are used which have received little or no processing then the traditional craft skills of chefs will be required to convert the food into attractive meals. At the other end of the spectrum, there are operations where very little traditional craft skills are required in order to finish the meal before serving it to a customer. This approach is typified by the fast food sector. However, this does not mean that the catering industry has been deskilled as the systems employed are increasingly technologically complex. Caterers, therefore, need new skills including an understanding of the processes employed throughout the food chain, and ways in which these processes affect the properties of foods. The education of caterers therefore needs to include science and technology covering the important basic and applied sciences which provide a foundation for modern catering. These include hygiene, food science, nutrition and the newly developed area of catering technology. As the growth areas in catering are those utilising pre-prepared products, the areas in decline are those in which the traditional craft skills play a dominant feature. Reliable statistics on the catering industry are few, however Table 1

TABLE 1
The Consumer Catering Market 1985–6

Sales through	£ million		%	
	1985	1986	1985	1986
Restaurants and cafes	1 970	2 070	26	25
Pubs	1 895	1 990	25	24
Fast food industry	845	1 000	11	12
Other take-aways[a]	1 280	1 345	17	16
Hotels	1 090	1 140	14	14
Retail outlets	220	230	3	3
Contractors	170	180	2	2
Clubs	150	160	2	2
Small cafes and others	80	85	1	1
TOTAL	7 700	8 200	100	100

Source: Euromonitor estimates.
[a]Mainly fish and chip shops and sandwich bars.

shows clearly the major growth areas in the fast food sector. The table does not include public sector catering, however this remains an important area approximately equal in size to the commercial sector in terms of numbers of meal occasions. Other areas undergoing expansion are pub catering and medium priced restaurants. There is very little growth in the high spend hotel and restaurant sector. The main common characteristic in the growth areas are their high dependence on technology in the operation of the catering system. As well as a mis-match in numbers there is also a mis-match in skills as most of the higher level courses are business and management orientated and have a small science/technology content. It is therefore important that catering courses provide a satisfactory grounding in science and technology alongside the more traditional catering subjects in order to ensure an educated workforce for the industry in the future. This problem is highlighted in the following review of catering education.

A recent study entitled *The Transfer of Technology in the Catering Industry* was sponsored by MAFF and carried out by the Hotel and Catering Research Centre, Huddersfield Polytechnic. Included in the study was an investigation into the science and technology content of current catering courses.[7] There are three types of catering courses designed for different levels of workers. These are:

City and Guilds — Craft/Operative
BTEC National Diploma — Supervisory
HND and Degree — Management

The emphasis in the past has been on craft training and higher level courses in catering have been slow to develop. A three year full-time National Diploma course was introduced in the 1960s and the BTEC National Diploma and the HND in the early 1970s. Catering degree courses began in the 1960s with two university degrees in Hotel and Catering Administration. Polytechnics have now developed a number of specialist catering degree courses, numbering 14 in total.

The study looked at all levels of courses. Examination of the syllabuses produced for City and Guilds courses and the BTEC Ordinary level courses revealed a satisfactory science input. The science appears in the guide syllabuses as food science encompassing microbiology, nutrition and relevant physics and chemistry. In the City and Guilds courses there have been two recent developments. First the development of a new syllabus for the basic cookery course, based on a number of taught competences. In this approach the basic cookery processes of catering, for example boiling, the use of microwaves and steaming, have been specified, and included in the specification is the relevant science and hygiene, microbiology, food science, nutrition, etc. It is envisaged that in each course the science is taught independently but reinforced in the catering practical and in the assessment of these competences. This approach is sound and is to be extended to other City and Guilds courses.

Another development from City and Guilds is the Background to Technology Project.[3] These are a collection of syllabuses which cover science projects needed to understand the technology used in a number of related industrial processes. Science themes form part of most City and Guild schemes either explicitly or implicitly. However, they are often 'hidden' and therefore candidates get little credit for the science learnt. Students who progress from one City and Guild scheme to another may also repeat the science already learnt. Therefore, students following City and Guild schemes do not always recognise how much science they have mastered or how well equipped they really are to cope with changes in technology, where the background science needed to understand the changes is fairly stable.

The Background to Technology Project has identified science topics from a variety of the Institute's schemes including catering and other related courses such as agriculture, food processing, bakery, etc. and produced a number of guide syllabuses. So far the syllabuses have been produced in physical sciences, chemistry and biological sciences. Development of the schemes only began in 1984 and therefore it is difficult as yet to comment on their utility. To date they have been used in schools where they have been offered as a science course to students intending to, or already following the Institute's Craft Schemes.

The Background to Technology Project is shown in the 706 scheme[1] as providing background science. It would be pleasing to see catering students taking parts of these certificates to have evidence of their competence in science. Its use in schools is also valuable in giving students confidence and competence in science prior to embarking on a vocational course of study. This initiative should, if successful, help to improve the application of science within catering courses and would help prospective students to appreciate in advance the scientific nature of catering.

The Business and Technician Education[4] Council administers two catering courses, the National Diploma in Hotel, Catering and

Institutional Operations (4 'O' Level entry) and the Higher National Diploma in Hotel, Catering and Institutional Management (1 'A' Level entry). For the lower level course the published core unit specifications include applied science encompassing food science, microbiology, nutrition, cleaning and energy conservation.

However, at a higher level the Higher National Diploma course does not include science in its published core unit specifications, indicating that science is not considered essential in the training of a catering manager. It is likely that most colleges do in fact include some scientific aspects in the training. However, the science may appear as science modules or be incorporated into catering lectures, therefore there can be no national standard. As the courses are business and management orientated, none contain in-depth scientific and technological knowledge. The science prescribed in the National Diploma core unit specifications would be adequate for management positions and should be included in the Higher National Diploma course as a minimum science input. This should include a study of the more technologically-based catering systems (e.g. cook–chill) and should be reinforced with relevant practical work such as equipment evaluation and recipe development. These aspects should be linked with the catering theory and practice. There should be links with the formal management strand, so that the concept of 'scientific management' can be introduced. This approach could include a study of the financial benefits of a cook–chill system, a planning exercise and consideration of staffing implications. The concept of Good Catering Practice (allied to Good Manufacturing Practice) could be promoted in this way as a vital management tool.

Therefore, both City and Guilds and BTEC Diploma courses provide adequate science cover in their schemes both in terms of content, background science and suggestions made for teaching approaches. Unfortunately, the translation of these schemes into actual taught courses is left to individual colleges. Doubtless some do the task very well, but the results are patchy and there is no national standard.

It has to be appreciated at the outset that students undertaking catering courses often have a poor background in science and may have chosen a catering course without realising the science input involved. Therefore, these students must be given confidence that they are able to cope intellectually with the necessary science. The science lectures must be able to capture and hold the interest of the students in the science subjects. This can be achieved in a number of ways, including teaching methods making use of practical sessions, preferably of the 'Nuffield' type and the development of strong links with catering practice — a team teaching approach.

Unfortunately the organisation in many colleges prevents these kind of activities from occurring. This is because the science is often taught by lecturers from a science department who have little knowledge of, or interest in, catering. These people find it very difficult to have any rapport with catering students. (Poor timetabling putting the science lectures on in the afternoon after a physically demanding catering practical does little to help the problem.) Another major problem is the lack of interest in science, or lack of scientific knowledge displayed by catering lecturers. This is a crucial factor, as the science must be reinforced by the catering lecturer. These lectures are respected by students and their involvement in a team teaching approach could greatly enhance the learning response. Unfortunately, many catering lecturers have poor science backgrounds themselves, having followed craft courses for their professional training. They lack the knowledge and the confidence to make the necessary science links in catering. Therefore a major problem of science teaching within catering courses appears to be the transfer of science to catering practice, by both the science and the catering lecturers.

Solutions to this problem are difficult to propose as they depend on initiatives within individual colleges. The adoption of a more integrated approach to science teaching is obviously a good start. Even better would be the appointment of scientists into catering departments as an integral part of the teaching team.

Examination of catering degree courses, of which there are currently 16 in the UK, shows that every one is individual and different. Most have some science content, but the depth and aspects covered are variable. However, all degree courses aiming to produce managers for an increasingly technologically complex industry should contain certain basic scientific aspects — food microbiology, food science and nutrition. The problem of transferring the scientific knowledge to catering practice should not occur at degree level, when the intellectual capabilities of the students enable them to make the necessary links. In addition, applied scientists such as food scientists and food technologists have been involved in the original concept and development of science-based degrees. These have been developed with the science and technology content implicit and providing core areas such as Catering

Technology and Catering Systems. The involvement of these scientists in catering education has had the effect of stimulating research into technical aspects of catering.

THE FIRST CATERING DEGREE COURSE AT HUDDERSFIELD POLYTECHNIC

The first science-based catering degree course began in 1971. It was the forerunner of the present BSc (Hons) Catering and Applied Nutrition run at Huddersfield Polytechnic. The degree is unique in its linking of catering with a specialised knowledge of nutrition and the applications of food technology. The objectives of the course are to produce scientifically educated catering managers able to manage technologically-based catering systems, apply the principles of nutrition to catering and provide a link between the catering and food manufacturing industries.

The course recognises the changes which are taking place within the catering industry, particularly the trend towards large centralised production methods made possible by new technology. The new technology involves processes such as chilling, freezing and the greater use of mechanisation in catering as well as developments in thermal processing and dehydration utilised by food manufacturers. These processes, as well as the traditional catering practices of cooking, warmholding and reheating, are gradually being put on a firmer scientific base. Alongside these developments there have been advances in nutritional knowledge and an increased awareness of nutritional issues by the general public. This is demonstrated by the interest created by recent expert reports on diet and health and the amount of media coverage given to issues on food, nutrition and health. There is now an increased demand for a healthier style of eating and one in which caterers have an important part to play. This has long been evident in hospital and social services catering, but now increasingly customers in commercial catering require meals based on sound nutritional principles.

All sectors of the industry, therefore, require managers able to analyse these complex needs, devise successful catering systems and lead an informed catering team and the degree aims to produce such graduates.

Course Organisation

The course is organised as a thin sandwich course containing two periods of supervised work experience. The first period serves as an opportunity for the student to apply his/her theoretical knowledge of catering to the practical situation. It also serves as an introduction to the supervisory and managerial roles of the caterer and is usually spent in a hospital, school meal or industrial catering situation. The second period may be undertaken in a variety of situations within the catering industry, the food manufacturing industry, the catering supplies industry and in research areas relevant to the course. The purpose of the second period of industrial experience varies with the placement but in a catering situation it is to extend the managerial responsibilities of the students. In the food manufacturing industry the students are introduced to quality control, production control and research and development techniques. This experience also serves to demonstrate the close links between the food manufacturing and catering industries.

Curriculum

The following subjects are studied in:
Year 1

Introduction to Catering: provides a foundation knowledge of the scale, structure and diversity of the catering industry and also gives practical and theoretical instruction in food production and service methods.

Food Legislation: provides an awareness of legislation and common law relating to food and catering.

Food Microbiology: deals with the characteristics of micro-organisms and emphasises the importance of food hygiene. The roles of micro-organisms in the spoilage and preservation of foods are also considered.

Financial Aspects of Catering Systems: introduces the basic financial aspects of accounting and cost control in catering organisations.

Foundation Sciences: provides a foundation knowledge of chemistry, biochemistry and physiology which serves as a basis for later subjects in the course.

Quantitative Analysis: provides a basic knowledge of mathematics, statistics and computing which can then be applied to other subjects in the course.

Years 2 and 3

Technical Aspects of Catering Systems: provides a detailed technical knowledge of catering processes, catering systems and all aspects of planning in a catering operation.

Food Science: deals with the composition, nutrient content, structure and properties of food commodities and the basic physical principles related to catering equipment.

Nutrition and Metabolism: introduces a study of nutrients including their dietary sources, function, effects of deficiency and biochemical involvement.

Social Aspects of Catering Systems: provides a social scientific knowledge of catering organisations, their workforce and customers.

Quantitative Analysis: develops the idea of experimental design and extends the ranges of statistical and computer analysis of data.

Year 4

Catering Systems: enables the use of General Systems Theory to be used as a means of analysing and synthesising the total catering system.

Catering Technology: involves an in-depth study of the science and technology of food processing as it relates to catering.

Applied Nutrition: examines the nutritional background to issues in catering including a detailed study of the relationships between diet and health.

Project: an in-depth study undertaken by each student of a topic which is of relevance to the course. Recent projects have included:

(i) Recipe developments for diabetics according to the latest dietary guidelines.

(ii) A comparative study of the use of convenience foods and fresh foods in composite dishes.

(iii) A systems analysis of a Travellers Fare operation.

(iv) An assessment of nutritional knowledge of a student population.

Graduate Employment

Graduates from the course have found employment in all sectors of the catering industry but especially in hospitals, school meals and social services, industrial catering, contract catering and the fast food sector. Further opportunities are offered by the food manufacturing industry. Positions held by graduates in this industry are involved with quality control, production control, research and development and marketing. More recently food retailing with the large supermarket chains has provided employment for graduates. Thus, the degree can be seen to be successful in providing scientifically educated graduates for at least three major parts of the food chain.

However, most managers in the catering industry today are craft trained or untrained and are ill-equipped to manage technical catering systems. Their inadequacies are manifest in the lack of scientific management techniques in practice. The absence of quality control, quality assurance and inadequate temperature control in the handling of foods, etc. are a few examples of these inadequacies caused by ill-defined policies and crisis management techniques. As these old style managers are replaced by graduates it is hoped that a more scientific approach to management will be developed.

As Higher National Diplomates and graduates attain full management positions in the industry this situation should improve, but only if their courses of study do include an adequate science/technology input. From this study it obviously depends on initiatives within individual colleges to develop satisfactory courses.

PART-TIME COURSES

In addition to the full-time courses discussed there are also part-time courses run for catering employees. These are usually City and Guilds basic course 705[2] and the more advanced 706,[1] aimed at personnel with no prior catering qualifications. Alongside these courses colleges often offer a basic course in food hygiene such as the Royal Society of Health's Food Handlers' course. Employees may also attend the newer courses offered by the Institute of Environmental Health Officers leading to the Basic, Intermediate and Advanced Food Hygiene Certificates. These courses are aimed at first tier workers, supervisors and supervisors/managers respectively.

The part-time courses presently offered are therefore primarily aimed at first tier workers and special emphasis is rightly given to hygiene. However, there is obviously a need to update current supervisors and managers in more specialised aspects. One of these must be technology to enable managers to make educated decisions in an increasingly complex and technologically-based industry. These courses could be offered by catering departments offering science-based degree courses in the form of taught masters degrees, post-graduate diplomas and specialist short courses.

Recommendations

1. The inclusion of science and technology in courses aimed at all levels of employees.
2. Acknowledgement by catering industry including its professional body, the HCIMA of the importance of science in catering courses.
3. More involvement by science teachers in the catering industry, in order for them to appreciate its technological nature.
4. The development of specialist courses on technological aspects of catering for existing supervisors and managers.

SOCIAL SERVICES CATERING

Social services catering has been selected for special attention in this paper for three reasons. Firstly it provides foods for a wide range of clients including a particularly vulnerable group in the population: the elderly. Secondly, its fragmentary nature and range of scale from domestic sized to large central production units, gives it problems whose solutions will only be achieved by a scientifically educated workforce. Thirdly, it is an area in which, to date, professional caterers have had very little involvement.

The term social services catering in this paper is taken to cover all local authority catering, excluding schoolmeals. The Meals-on-Wheels service alone provided 30 million meals in 1985 to recipients in their own homes. In total these authorities provide 43 million meals at an approximate cost of £46 million, or just over £1 per meal. The fragmentary nature of the catering services may be due to the fact that some of the Acts of Parliament concerning the social services were permissive. Permissive Acts are those which are open to interpretation and which are implemented and developed at the discretion of the individual local authorities.

Many of the catering services are provided in residential homes and day centres which are overseen by general managers who have little specialist catering expertise. In addition, the catering policies are often embodied in the concept of 'total care' leading to a wide interpretation. In some authorities, these factors lead to a lack of definitive catering policy. This in turn creates difficulties in identifying the specific aims and objectives necessary to run effective catering services for specific groups of people. This lack of definitive policy and apparent low priority of catering services is reflected in the fact that some local authorities still employ no professional manager or adviser for this service. The difficulties so created may be summed up by quoting Mr M. W. Bishop, Director of Social Services in the County of Cleveland (in 1987). He stated that 'officers in charge (of residential homes and day centres) do not have the skills and abilities to complete the role of caterer, let alone the time'. He further stated that 'there is a need for catering officers, either as managers or advisers, centralized in order to obtain standards throughout a local authority'. He defined the role of the caterer as 'manager, dietary expert, standards developer, budget controller, purchaser, nutritional expert and trainer'. His description certainly does not portray the role of the amateur. Indeed, dedicated local authority caterers in 1985 formed the Advisory Body for Social Services Caterers (ABSSC). The ABSSC are fully aware of the current problem including, in many areas, the shortage of professional expertise. They intend to function as a 'self help' group.

This body proposes to collate the collective work of regional groups, produce policy recommendations and collectively discuss relevant legislation, reports and guidelines and their impact on social services catering. They intend to produce a *Social Services Catering Manual* as a reference for all Social Services Departments, and publicise their work via the press and an annual seminar. A major aspect of their collective work will be to prepare and issue an annual report to their Directors of Social Services.

The perception of the importance and need for professionalism was reinforced in 1988 by the results of a questionnaire carried out (by M. Malik) with the co-operation of 14 local authority officers responsible for catering services in either an operational, advisory or administrative capacity. They identified 12 major groups of clients requiring catering services, most of whom fell into vulnerable categories. These groups of clients are shown in Table 2.

However, of the 14 respondents, only 10 stated that they were professionally qualified caterers employed to manage, advise, develop and control the services. Three respondents had similar responsibilities but had no specialist qualifications and one respondent described his/her status as amateur.

The major skills identified as important in catering education included interviewing, selection and training techniques, and the

TABLE 2
Major Client Groups in Social Services Catering

Elderly
Meals-on-Wheels
Residential Homes
Day Centres
Luncheon Clubs

Children
Residential Homes for the Mentally Handicapped
Residential Homes for the Physically Handicapped
Residential Homes for Children in Care

Others
Day Centres for Physically Handicapped Adults
Day Centres for Mentally Ill Adults
Residential Homes for Physically Handicapped Adults
Residential Homes for Mentally Handicapped Adults
Social Education Centres

management of staff and organisations. Required operational techniques included budgeting, financial control, purchasing and stock control. It was emphasised that an appreciation of catering in the context of social services including an understanding of the clients social and nutritional needs was essential, as was the ability to research and transfer information on current catering systems and technologies. Communication skills are also required in the education of senior administrative and executive personnel in order to achieve a professional catering service.

Only two of the respondents felt that currently available advanced courses equipped managers to deal with the complexities of social services catering. The major omissions were identified as food hygiene, the understanding of the scale and complexity of social services catering operations, large scale production techniques, including kitchen organisation, and a knowledge of local authority funding. Only five respondents stated that all cooks and food handlers were required to have at least basic craft qualifications. Of the nine services not requiring these qualifications, four offered pre-service and five offered in-service training. The standards of operational skills were monitored in a variety of ways, ranging from random checks to formal evaluation by the training section. Only three of the respondents were satisfied with the level of available operational skills and the steps being considered to remedy this situation included on-the-job training, training of unit general managers, day release courses and the employment of qualified cooks only. Attempts were being made to influence the directors and Social Services Committees (presumably to understand the importance of catering and the need to upgrade skill levels in order to improve the service overall).

The inhibiting factors for the implementation of upgrading programmes included lack of time and funds, lack of training structures and trainers, and the unavailability of 'tailor made' courses. Poor staff attitudes to training were also cited as a problem. It was also felt that there was a lack of recognition by the local authorities of the need for a training initiative. This reflected some authorities' apparent disinterest in the service and their designating it as only a low status function.

From the authors' personal knowledge of the services, the picture drawn by the 14 respondents is representative of many others throughout the country. It is a picture of valiant efforts by too few professionals attempting to cope with enormous and changing demands with limited resources which includes human resources.

This short survey also appeared to confirm the prediction of the Education and Training Advisory Council[9] that there will be an increasing gap between the supply and demand of professionally qualified personnel as the number of graduates and diplomates remain relatively static as the industry continues to expand.

CATERING IN THE FUTURE

Thus, in summary, from the example taken of social services catering, it can be seen that the catering industry requires managers able to analyse and take decisions on complex situations. Some of the issues likely to face catering management in the future have been discussed at a recent conference called 'The Future of Technology in Catering' held by the Hotel and Catering Research Centre of Huddersfield Polytechnic and sponsored by MAFF.[6] The conference gathered together invited caterers, food manufacturers, catering equipment manufacturers, food scientists, nutritionists and environmental health officers to discuss the future directions of the catering industry. The conference's main predictions were:

1. A continued blurring of the commercial and non-commercial sectors.

2. An accelerated trend towards the decoupling of food production from food service caused either by caterers operating central production units or by food manufacturers supplying caterers with ready-prepared meals.

3. Consumer demands will increasingly expect safer food with more nutritional information and fewer additives.

4. Changes in food supplies will include new varieties of agricultural products, some from the results of biotechnology, and an increasing emphasis on vegetables, fruits, cereals and seafood linked with continued trends towards healthy eating.

5. Moves towards designing out of hygiene problems by the use of processed foods typified by the fast food sector in which foods requiring only fast reheating before service to the customer are employed, therefore involving little preparation and therefore little time for hygiene problems to develop.

Recommendations for the future

1. *Hygiene Risks*

   Sectors such as hospitals and restaurants continue to have hygiene problems. These will only be controlled by staff training and a well documented food control procedure.

2. *Time–Temperature Control*

   This is essential in catering, especially in cook–chill catering operations. Control of time and temperature is vital as it can affect bacterial growth, nutritional content and sensory qualities of foods. Poor control of time and temperature causes food poisoning. Temperatures should be monitored and recorded in the key operations of cooking, reheating and chilling.

3. *Catering Education*

   In view of these predictions and recommendations the report concluded that with the move towards more complex technology, the problems of hygiene, the influence of nutrition through healthy eating and the need for good scientific quality assurance, all catering management courses should include a strong scientific and technological element.

SUMMARY

The catering sector has been designated for expansion by the National Advisory Body but this will not be possible unless more young people choose a career in the industry. There are now fewer eighteen-year-olds available to enter courses so it will be difficult to increase intakes unless schools give better guidance on careers in catering. It must be recognised that many aspects of catering require a scientific education. Science sixth formers need to be informed of catering as a respectable career opportunity. Seventy-five per cent of students following diploma and degree courses are female. We therefore need more males in catering education. However, even with expansion, it will take a long time for colleges to provide the industry with an educated workforce. So, in the meantime, colleges and industry need to develop closer links and to work together to develop up-dated courses for existing supervisors and managers. No improvements can be expected unless the industry has better educated managers. The industry has been reactive for too long; improving the educational status of caterers should bring about a more pro-active workforce able to play a unique role in the food chain, due to its direct link with the consumer.

REFERENCES

1. City and Guilds 706 *Cookery for the Catering Industry 1986–87.*
2. City and Guilds 705 *General Catering Certificate (New Scheme) 1985.*
3. City and Guilds *Background to Technology Project 366,* Pilot Scheme 1984.
4. Business and Technician Education Council (1987) *Hotel, Catering and Institutional Operations. Care Unit Specifications and Sample Learning Activities.*
5. The Institution of Environmental Health Officers (1986) *Basic, Intermediate and Advanced Food Hygiene Certificates. Examination Regulations and Syllabus.*
6. Hotel and Catering Research Centre (1987) *The Future of Technology in Catering,* Huddersfield Polytechnic.
7. Hotel and Catering Research Centre (1987) *The Transfer of Technology in the Catering Industry,* R. Collison and A. West (Eds), Huddersfield Polytechnic.
8. Gallup (1985) *British Survey of Eating Out.*
9. Hotel and Catering Industry Training Board (1983) *Hotel and Catering Skills — Now and in the Future,* Report commissioned by Education and Training Advisory Council.
10. HCIMA (1987) Survey of enrolments for session 1987/88 degree, diploma and higher diploma courses in Hotel Catering and Institutional Managements. Hotel Catering and Institutional Management Association, London.

'Where do Carrots Grow?' —
Ideas for Teaching and Learning about
Food and Diet in Schools

Sheila A. Turner

Department of Science Education, University of London Institute of Education,
UK

ABSTRACT

This paper looks at the following issues in relation to food, diet and education in schools:

1. *The implications of a national curriculum for teaching.*
2. *The choice of appropriate topics for pupils in primary and secondary schools.*
3. *Implications for teachers of new courses including GCSE, TVEI and CPVE.*

The publication of the National Curriculum discussion document by the DES in July 1987 has caused all those involved in education to reassess the content of the school curriculum. The document makes explicit the division of the timetable into discrete units with particular importance being given to language, mathematics and science. It will be argued that teaching about topics such as food production and food processing should involve interdisciplinary studies; the future of such studies as part of a national curriculum has yet to be determined.

The choice of topics for inclusion at primary and secondary school level depends on many factors. Different approaches to teaching and learning about food and diet influence learning outcomes. Examples of topics and the ways in which they can be taught to pupils of different ages are outlined.

143

In primary schools food production may be covered as part of themes or topics such as 'Food around the world' or 'The market'. Such topics will normally include aspects of geography and history plus, in some instances, science and technology and mathematics. The constraints of 'specialist' timetabling in secondary schools tend to shift studies of food production into particular subject areas such as geography. Work on diet and health may feature as part of a number of subjects including home economics and biology. The broadening of many geography, home economics and science courses for secondary pupils in the past two years has led to the development of new resource materials.

The proposals for a national curriculum and new courses, such as GCSE or CPVE, have implications for all areas of the curriculum and for the resources which are needed for schools. The types of resources which may be required, from books to in-service education for teachers, are examined.

INTRODUCTION

In this paper I want to look at three main areas in relation to food production and education in schools, namely:

1. The implications of a national curriculum.
2. The choice of appropriate topics for pupils in primary and secondary schools.
3. The implications for teachers of new courses including GCSE, TVEI and CPVE.

THE NATIONAL CURRICULUM

At the present time any discussion about education in schools must start with an examination of the government proposals for a national curriculum as outlined in the Education Reform Bill (1987) and in the DES discussion document published in July 1987.[1] The proposals have profound implications for the way in which schools are governed and managed and for the organisation and content of the school curriculum.

The National Curriculum document makes explicit the division of the curriculum into discrete units with particular importance being given to language, mathematics and science for all pupils from 5–16 years. In primary schools most time will be devoted to these three core

TABLE 1
Possible Allocation of Curriculum Time in Years 4 and 5 in Secondary Schools

Foundation subjects	Time spent on foundation subjects (%)[a]	Additional subjects e.g. for GCSE might include:	
English	10	Science	
Maths	10	Second Modern	
Combined Sciences	10–20	Foreign Language	
Technology	10	Classics	
Modern Foreign		Home Economics	
Language	10	History	10%
History/Geography *or*		Geography	
History *or* Geography	10	Business Studies	
Art/Music/Drama/Design	10	Art	
Physical Education	5	Music	
		Drama	
		Religious Studies	

[a]These percentages do not appear in the Education Reform Bill.
Source: DES — *The National Curriculum 5-16: A Discussion Document* — July 1987.

subjects. All pupils will also study a number of 'foundation' subjects, these include technology, history and geography, plus a number of additional subjects. One suggested allocation of time for Years 4 and 5 in secondary schools is shown in Table 1. Home economics, a subject in which much nutrition education occurs, is not a foundation subject. Although its place as a valuable part of the curriculum is acknowledged by the Department of Education and Science[1] its future in the school curriculum will be determined by Local Education Authorities and governing bodies of schools. As is apparent from Table 1 the DES does not recognise health education as a separate subject. It suggests that health education can be taught through foundation subjects such as biology.

The emphasis given to science and technology by the DES and the lesser weight given to home economics and health education is of particular relevance in the context of topics such as food production and aspects of health education. It is easy to be pessimistic about the 'marginalisation' of home economics in the national curriculum proposals. I think that we need to consider the topics which have traditionally been taught in home economics in the wider context of the whole curriculum. We are being presented with a valuable opportunity to reconsider the ways in which we teach food-related topics — a process which has already begun in the development of new GCSE syllabuses and meetings such as that held in Bangalore in 1985[2] which have looked at ways in which the traditional subject boundaries could be broken down. Furthermore the case for balance in the curriculum as outlined by the DES in *Better Schools*[3] needs emphasising. The DES states that each part of the curriculum 'should be allotted sufficient time to make its special contribution, but not so much time that it squeezes out other essential parts'.

The view of science education implicit in the national curriculum proposals is very similar to suggestions made by the DES in *Science 5–16: a statement of policy.*[4] The policy document stresses the need for science teaching to have breadth — it should include for example the study of technological applications and the social consequences of science — and relevance, it should also make links across the curriculum. In primary schools it is usual for work in science and technology to be linked with other components of the curriculum. Such links are rare in secondary schools. Subject specialists, for example in science and home economics, need to work more closely together if the contribution that each can make to the other is to be fully exploited. In the past it has been too easy for the curriculum of intent to be divorced from the curriculum of reality. It is to be hoped that the national curriculum proposals will facilitate the implementation of much needed interdisciplinary studies in secondary schools. It is encouraging that the interim report by the National Curriculum Science Working Group[5] has indicated that it is planning, as part of its next phase of consultation and discussion, to look at links between science and other subjects and subject areas: home economics and health education are both mentioned in this context (paragraph 36). Such links across the curriculum are in my opinion essential if teaching about 'The Human Food Chain' is to be successful.

The national curriculum proposals state that a significant proportion of time — perhaps 10% — should be devoted to technology. It would be possible for home economics to make an important contribution to the curriculum through this foundation subject. Aspects of food science, including biotechnology, could be incorporated readily into broad-based technology, as well as science, courses. The modular structure which is becoming increasingly adopted for GCSE, for example in science, is another mechanism by which the valuable contribution made by home economics to nutrition education could be utilised.

The fate of health education, including aspects of nutrition education, hangs in the balance. The brief mention of health education in the National Curriculum consultation document[1] is hardly reassuring for those of us who have urged its inclusion as an essential component of the curriculum. The final report of the National Curriculum Science Working Group in June 1988 will perhaps help to establish some of the contexts in which health education can contribute to science education. If the interdisciplinary links and broad-based, relevant courses advocated above are implemented then it is possible that teaching about diet and health will feature as part of courses for all pupils, something which does not happen at present. Many schools, especially primary schools, already have policies for Health Education. Such policies need to be recognised as important by all those involved in education, by teachers, by parents, by governing bodies, by LEAs, by the DES, if their objectives are to be realised.

TEACHING ABOUT FOOD AND DIET

Ideas about Teaching and Learning

Ideas about teaching and learning are always subject to change and debate. Much current research in science education is focussing on the ideas and understandings which children already have and the ways in which these ideas — often referred to as alternative frameworks — affect the ways in which children learn.[6] The findings from such research indicate how important it is for teachers to take account of the understandings which children already have when planning teaching schemes. The Children's Learning in Science Project (CLISP) based at the University of Leeds is currently investigating teaching strategies which could be used to help children's understanding of scientific concepts; their report on teaching about energy[7] has implications for teaching in the whole curriculum. They suggest that teaching/learning programmes should start with the ideas and skills which pupils bring to the learning situation. Where necessary these ideas can be developed, extended and changed; this may be part of a long term process of development. The contexts for learning should be selected to be relevant to a pupil's life and the teaching methods designed to give pupils opportunities to construct, exchange and evaluate ideas. Pupils should be actively involved in their own learning and take responsibility for it.

Such teaching/learning situations place emphasis on small group discussion and feedback, with the teacher as a manager of the learning environment rather than as a provider of knowledge. The understanding of concepts which accrues should, it is hoped, enable pupils to apply what they have learnt in everyday life.

At the present time knowledge about children's ideas about food, including why we need food and where it comes from, is limited. What is apparent is that the way in which children and adolescents make decisions about what they will, or will not, eat is a complex process which is influenced by many factors including those shown in Fig. 1. Part of our job as educators is to help children to become aware of the reasons for the ways in which they make choices and decisions about their diet. Teaching of this type involves utilising a range of teaching strategies and providing the type of learning environment outlined by CLISP.[7]

If children are to apply what they have learnt about food and diet in everyday life we have to help them to develop personal and social awareness and decision-making skills. The development of such skills is not an easy task; it is being attempted in the new Technical and Vocational Education Initiative (TVEI) described later in this paper.

The importance of providing opportunities for pupils of all ages to identify with real situations and events needs general recognition as does helping children to find solutions to problems and questions which they themselves have posed. Such aims underpin the reports on *Better Science* produced by the Secondary Science Curriculum Review,[8] including that on Health Education, which indicate ways in which such approaches can be included in classroom situations.

Topics Related to Food and Diet

The choice of topics for inclusion at primary and secondary school level will depend on many factors including the age, ability and background experience of the pupils as well as 'external' factors such as examination syllabuses. The suggestions for teaching outlined below provide examples of ways in which studies on food and diet can be developed using teaching strategies consistent with the views of pupil learning discussed earlier. Topics are described therefore in terms of processes and contexts as well as content.

In primary schools food production may be covered as part of themes or topics such as 'Food around the world' or 'The market'. Such topics will normally include aspects of geography and history plus, in some

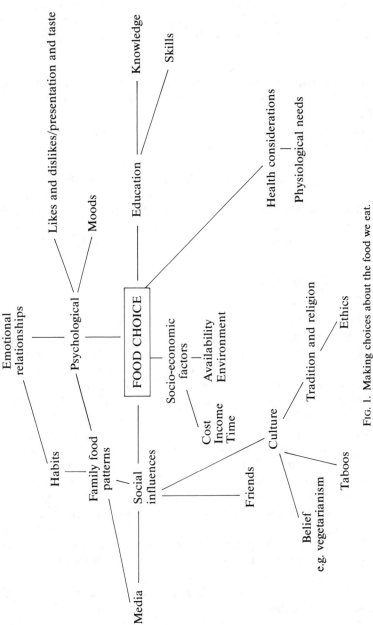

FIG. 1. Making choices about the food we eat.

instances, science and technology and mathematics and may involve studies over a number of weeks. Figures 2 and 3 illustrate how teaching about the market or specific foods could be developed in different areas of the curriculum.

The starting point for work on a food topic with children of all ages could be a visit to a local market or farm. For very young children a visit to a market might have the specific purpose of buying particular types of fruit or vegetables; purchasing exact numbers or amounts gives valuable experience of handling money, counting and estimating. The fruit or vegetables bought could be used as the basis for a 'snack' at school, encouraging children to try foods they have not sampled before. The snack prepared might involve cooking some of the items bought, for example, using vegetables to make soup. Alternatively the food could be used for close observation work based on drawing, painting or printing. A whole range of scientific investigations could be developed involving the senses, including taste and smell. Questions asked by children such as

— are carrots crunchy?
— how many people in our class like carrots?

could also be the start of investigations. The art of asking questions is a very important aspect of education and one of the skills which has been identified as fundamental in science education.[5] The questions which children ask can also alert teachers to misunderstandings or worries which children have about aspects of food and diet.

Older pupils in primary schools might use a visit to the market for surveys to find out more about what sorts of food people buy or to investigate where different foods come from. Such surveys could be used to initiate studies of food sources, distribution and marketing. Longer term surveys could enable children to learn more about seasonal variation including the availability and varying price of different foods. Questions such as 'how are tomatoes packed to stop them being damaged on the way to the market?' can lead to a range of problem solving activities linked to different types of food packaging.

The constraints of 'specialist' timetabling in secondary schools tend to shift studies of food into particular subject areas with teachers of those subjects often being unaware of what is going on in other departments. Work on diet and health may feature as part of a number of subjects including home economics and biology; food production may be studied in geography. There is evidence that most nutrition

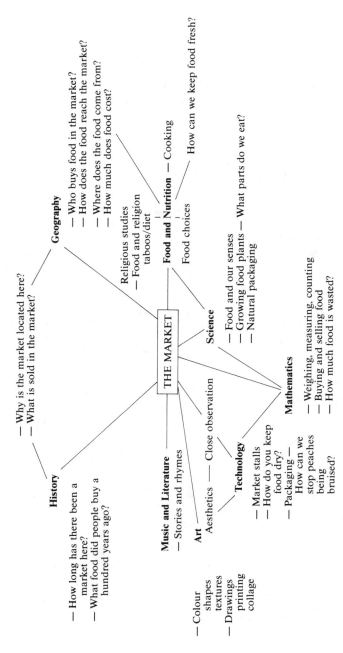

Fig. 2. The market-place as a starting point for teaching about food and diet in primary schools.

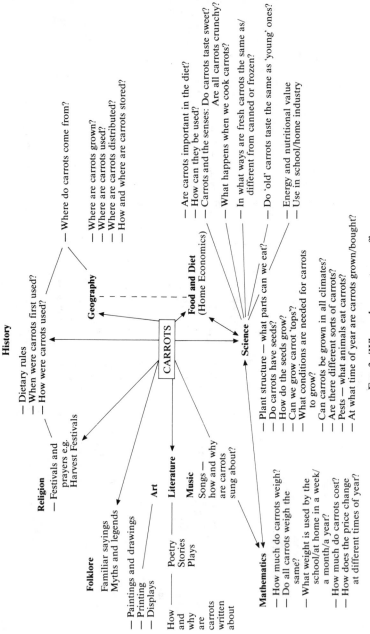

FIG. 3. 'Where do carrots grow?'

education takes place in biology and home economics.[9] The importance of making links across the curriculum when teaching about food and diet has already been mentioned. The curriculum links shown in Fig. 3 could be implemented at secondary level in the same way as normally occurs in primary schools. There are important advantages in teachers from different disciplines learning from each other: humanities departments often utilise teaching strategies, such as role play or simulations, which could be of considerable value to teachers in other departments.

The introduction of General Certificate of Secondary Education (GCSE) courses has led to a change of emphasis in many areas of the curriculum for pupils of 14–16 years. There has been a welcome reduction of traditional content and a move towards more skills-based courses. The emphasis on relevance and applications in new syllabuses is also important in the context of making links across the curriculum; it may ultimately enable the separation between traditional areas of the curriculum to be reduced.

Teaching Materials

The introduction of new courses generally coincides, not unsurprisingly, with the development of new resource materials. Some of the materials, including books, developed in the past three years reflect current thinking about teaching and learning and are designed to develop a range of skills and attitudes including those listed below for science:[5]

Skills
- observing
- raising questions
- hypothesising
- classifying
- predicting
- testing
- planning and carrying out investigations
- communicating
- measuring
- interpreting data

Attitudes
- curiosity
- perseverance
- creativity/inventiveness

— open-mindedness/flexibility
— critical thinking/reflection
— honesty
— co-operation with others
— willingness to tolerate uncertainty
— respect for living organisms and the environment
— appreciation of the relevance of science and technology to everyday life
— awareness that the applications of science and technology have social implications

Most of the skills and attitudes listed here have been a part of educational thinking for some time, particularly in primary schools and health education courses, most are applicable across the curriculum. The types of materials developed in the past two years for 11–13 year olds in science, for example in *Warwick Process Science*[10] or ILEA *Science in Process*[11] reflect a skills-based approach to teaching and learning which is similar to primary practice and underpins GCSE courses.

The same type of approach is exemplified in the ILEA *Nutrition Guidelines*[12] with its emphasis on skills development and socio-economic dimensions of nutrition education including food production and food marketing. Although the guidelines are aimed at home economics teachers they also have relevance across the curriculum.

The move to broader based courses which take account of social, economic and environmental factors is reflected in materials designed to be used as adjuncts to existing courses. An example is the *Science and Technology in Society*[13] teacher and pupil booklets designed to help pupils of 14–16 years to learn about the applications of science in the 'real world'. A wide range of learning strategies can be utilised, including discussion, role play and problem solving. Significantly the materials are proving of value in teaching subjects other than science and are a further example of how meaningful cross-curricular links can be made. Topics include study of aspects of food production, food labelling and food prices as well as fibre in the diet and making fertilisers.

NEW COURSES

For those who are not working in schools the education 'jargon' of the past decade — including GCSE, TVEI and CPVE — must be bewildering.

Teachers are themselves still coming to terms with new examinations, course structures and curriculum initiatives. The review of new courses in this section examines the implications for teaching of three recent initiatives, the General Certificate for Secondary Education (GCSE), the Technical and Vocational Education Initiative (TVEI) and the Certificate of Prevocational Education (CPVE).

The General Certificate for Secondary Education (GCSE) has received considerable publicity in recent months. In June 1988 the first cohort of pupils will sit examination papers for the new GCSE which has replaced 'O' level GCE and CSE examinations. Attention has focussed on coursework assignments, which form part of the assessment, and teacher assessment. A more significant factor, which receives less publicity, is the fact that all examination syllabuses have now to be approved by the Secondary Examinations Council (SEC): to be approved syllabuses have to meet national criteria which specify assessment objectives for examinations. In addition to general criteria there are subject specific criteria which cover areas such as core content and the relationship between the core content and assessment objectives. GCSE syllabuses, which place emphasis on the development of understanding and skills, have to be designed to help pupils to understand the relationship of the subject to other areas of study as well as its relevance to their own lives.[14] Stress is also placed on the development of awareness of economic, political, social and environmental factors. In science one of the aims is to enable pupils to acquire sufficient knowledge and understanding so that they can become 'confident citizens in a technological world, able to take or develop an informed interest in matters of scientific import'.[14] If these aims are achieved young people should leave school informed about general societal issues and with concomitant personal skills which will enable them, for example, to ask questions about the 'Food Chain'.

Although all Examination Groups have developed a range of different syllabuses for GCSE, schools are also able to develop their own syllabuses. Some schools have taken the opportunity to design courses which they see as more suitable and relevant for their pupils. Such syllabuses, allied to the more general adoption of modular courses, provide further opportunities for teaching about food in ways which permit coverage of aspects which have been ignored in the past.

The introduction of TVEI in 1983 was of the utmost significance for education. For the first time the initiative and funding came not from the Department of Education and Science (DES) but from the

Manpower Services Commission (MSC). Both the funding arrangements and the nature of the proposed courses worried many educators. TVEI is designed as a four-year programme for 14–18-year-olds which will provide full-time technical and vocational education, with appropriate work experience, in conjuction with academic courses. The courses should cater for pupils of all abilities, including the most able. It is hoped that pupils who have been involved in TVEI will be able to make more realistic choices about jobs and careers.

Each pupil follows a core curriculum which includes information technology and career and vocational guidance and counselling. Emphasis is placed on skills including communication, numeracy, problem solving and personal and social development. There is also opportunity for pupils to choose from a range of courses. Some schools offer GCSE courses in less traditional subject areas as part of TVEI such as 'agriculture, horticulture and plant husbandry' or catering. One of the interesting outcomes of TVEI has been the variety of courses and initiatives upon which schools have embarked. One school, for example, has smallholdings managed by pupils which supply goods to local hotels. In another school the Home Economics and Rural Science Departments have joined together to produce and market a variety of items including yoghurt and herbs.[15]

It is still too early to assess the impact of TVEI although a recently published report based on a study of twelve schools[16] indicates what has been achieved. The researchers found positive developments in some schools including consultative team-teaching. One of the findings is that it is difficult to develop innovative courses where examination boards are not geared to examine cross-curricular activities. However desirable such subjects may be in educational terms they are regarded as having low status by pupils, parents and employers if they are not examined. The researchers also suggest that it will take time for most teachers to adapt to the new styles of teaching implicit in TVEI provision.

In spite of early misgivings it is obvious that TVEI has much to offer pupils but its ultimate success will depend on genuine commitment on the part of schools and the community. Such commitment includes the necessary funding for training, resources and research into the types of courses which will be of most benefit to young people. The types of cross-curricular links advocated earlier in this paper are still far from common.

The Certificate of Prevocational Education (CPVE) is similar in some respects to TVEI but provides a one year course for students of 16–18 years who are not following 'A' level courses in the sixth form. Like TVEI it includes work experience and a core of studies which focusses on the development of skills such as communication and problem solving. The core also includes industrial, social and environmental studies plus science and technology. Work is assessed typically by means of assignments or mini-projects. Students also take vocational studies which are designed to introduce them to employment: these studies are composed of five categories which include production and distribution.

The design of CPVE is such that traditional subject boundaries should be eroded. Furthermore, in many instances teachers are involved in co-operative teaching and teaching in disciplines other than their own. Some staff are also closely engaged in working with local employers. There are many positive gains in such links with employers, for example learning about aspects of the 'Human Food Chain' with which teachers may not be familiar. Experience gained by teachers through their involvement in CPVE may also give them the confidence to develop closer cross-curriculum links when teaching younger pupils. It should add an extra dimension to the content and approach taken in teaching about particular topics.

DISCUSSION

The proposals for a national curriculum and new programmes of study such as GCSE or TVEI, have implications which extend beyond schools. There are concerns that these initiatives will be successful only if sufficient resources are made available by government; resources include particularly, numbers of specialist staff and teaching materials, such as textbooks and laboratory equipment. Furthermore, altering the curriculum has little effect unless teachers are able to modify their teaching strategies. Changes in teaching strategies occur only when teachers feel confident and secure in the level of support which is available. Support includes the provision of adequate funding for specialist teachers, resource materials and in-service courses (INSET).

The funding of INSET has changed fundamentally in the past year with the introduction of GRIST (Grant related in-service training) which gives LEAs discretion in the way in which money for INSET is used locally. Some LEAs have delegated much responsibility for INSET to individual schools. The autonomy given to schools for planning their own in-service programmes in response to felt needs, including implementation of a national curriculum, may be one of the positive benefits of the new system. On the other hand it may be that GRIST will restrict the type and nature of in-service provision; priority may be given to local and short term goals which may not accord with long term or national needs. Although provision for science and technology remains a high priority GRIST may make it more difficult for effective initiatives to take place in these subjects. Both the provision of resources and the nature of INSET have implications for the types of courses which have been identified as important in teaching about the 'Human Food Chain'.

The changes which have taken place in schools in the past ten years reflect as always a mixture of political expediency, economic constraints and changing ideas regarding the ways in which children learn. We have a long way to go before young people leave school with all the knowledge and skills that might be deemed desirable. Progress, albeit slow progress, is being made however, and if we as educators can achieve the aims outlined earlier in this paper then school leavers will be equipped with the skills necessary to take their place as confident, open-minded, questioning citizens who will find it possible to forge the links in the 'Human Food Chain'.

REFERENCES

1. DES (1987) *The National Curriculum: a discussion document,* Department of Education and Science and Welsh Office, London.
2. Lewis, J. L. and Kelly, P. J. (Eds) (1987) *Science and Technology Education and Future Human Needs,* Pergamon Press, London.
3. DES (1985) *Better Schools: a summary,* Department of Education and Science and Welsh Office, London.
4. DES (1985) *Science 5–16: a statement of policy,* HMSO, London.
5. DES (1987) *National Curriculum Science Working Group: Interim Report,* Department of Education and Science and Welsh Office, London.
6. Osborne, R. and Freyberg, P. (1985) *Learning in Science — the implications of children's science,* Heinemann, London.
7. Driver, R., Wells, P. and Brook, A. (1987) Children's understandings and the teaching of energy in schools. In *Energy in the School Curriculum, 1987 Conference Report,* British Gas.
8. Secondary Science Curriculum Review (1987) *Better Science — Curriculum Guides,* Heinemann Educational Books/Association for Science Education for School Curriculum Development Committee, London.

9. Wilcox, B. and Gillies, P. (1983) The co-ordination of secondary school health education. *Education Research,* **25** (2), 98–104.
10. Screen, P. (1987) *Warwick Process Science,* Ashford Press Publishing, Southampton.
11. ILEA (1987) *Science in Process,* Heinemann Educational Books in association with the Inner London Education Authority.
12. ILEA (1985) *Nutrition Guidelines,* Heinemann Educational Books, London.
13. Association for Science Education (1986) *Science and Technology in Society,* ASE, Hatfield.
14. Secondary Examinations Council (1986) *Science — GCSE a Guide for Teachers,* Open University Press, Milton Keynes.
15. Technical and Vocational Initiative (1988) *TVEI — Insight* No. 12, London.
16. Technical and Vocational Initiative (1988) *The TVEI Curriculum: 14–16,* TVEI, London.

Education about Infant Feeding

PETRONELLA CLARKE

Department of Health, London, UK

ABSTRACT

Several aspects of infant feeding need to be included in an educational programme. Mothers will first make the choice of whether to breastfeed or not. Once that decision has been made, education about how to feed by the chosen method is called for. As the baby grows parents will want to know when to diversify the diet and, later, what diet will best safeguard their baby's mental and physical development and health.

There is an undoubted need for well-informed education about infant feeding. Infants are vulnerable and dependent on their care-takers. Parents are eager for information and, unless correct and attractive sources are available, they will accept whatever sounds authoritative to them especially from family and friends.

Many of the factors influencing decisions about infant feeding are attitudinal. If education is to influence parents' behaviour it will need to be planned on a long time scale. At all stages the education must seek to support the parents' confidence, not sap it.

The mainstay for education in infant feeding of parents and other care-takers is the health professionals, especially midwives and health visitors. Health educationalists and informed lay groups also make a contribution. It follows that the primary medium for education is personal discussion. Booklets, magazines, books, recordings and film can provide endorsement and support for the health professional's message.

INTRODUCTION

Most children will, at some time, have demanded to know how to feed a baby blackbird or hedgehog found in the garden. They may be less eager to know how to feed a human baby. Nevertheless, the moment will come, when these children are themselves parents and face the challenge of providing for their own child. Infant feeding becomes of central concern and the basis for the very survival of their newborn. Most parents perceive their baby as extremely vulnerable, often more so than in reality, and there must be few parents who have not had pangs of worry about whether they can cope with feeding their baby. Even parents who are fully informed about infant feeding will welcome support and encouragement in translating theory into practice.[1]

Several features are unique to education about infant feeding:

(a) Breastfeeding is a personal activity which is closely connected with the mother's self awareness and sexual identity. These are aspects of personality which are linked to attitudinal factors.[2-4]

(b) The baby's diet has to change. The first year of life exceeds all others as a period of maturing, learning and growing. New mothers may just have become confident with one feeding schedule when the baby's needs outstrip what is being provided both in amount and type of food.

(c) Parents want to know how to plan a diet which ensures their children's future health both in the short and in the long term. They want to know when and how to introduce a mixed diet. What are the special needs for my baby during these periods of very rapid growth? Should their child's diet be moved in the direction of so called 'healthy eating' as recommended for adults?

Parents need guidance which is authoritative and which can be justified on commonsense grounds and in the manifest wellbeing of their baby. In the best circumstances this education will also lead to an enhancement of the parents' confidence in nurturing their child.[5]

EDUCATION ABOUT THE CHOICE — TO BREASTFEED OR NOT

The Nutritional View

The newborn needs food which can be started immediately after birth and which is compositionally adequate and microbiologically safe. Breastfeeding fulfils these requirements and has the safeguard of a history of millions of years of success. Breastfeeding requires the minimum of prior education and, once established, most mothers find that breastfeeding also expresses their desire to nurture their precious baby.

Alternatives to breastfeeding cannot match the history of this success. However, insofar as clinical and technical knowledge is available, artificial feeds can provide an alternative to breastfeeding in Britain.[6] The manufacture and preparation of artificial feeds is prey to human error especially in less developed countries where babies may be placed at risk if they are fed infant formula which has not been prepared correctly.[7]

Representative surveys of infant feeding practices in England and Wales done by the Office of Population, Censuses and Surveys (OPCS) have shown that in 1975, 51% of infants were put to the breast at birth, this figure was 67% in 1980[8] and 65% in 1985. (Table 1).

When will Education have its Greatest Impact?

By the time of the birth 94% of women have decided how they intend to feed their babies.[3] The timing of mothers' choices about whether to breastfeed suggests that education is specially valuable before and during the first pregnancy. Thirty per cent of mothers of first babies in the 1975 OPCS survey of infant feeding in England and Wales said that they had decided how they would feed their baby before they became pregnant.[2] For second and later births the method of feeding an earlier child is a key determinant in the mother's decision for her present pregnancy. The scope for education during a first pregnancy will be limited if the mother is preoccupied with the impending labour and

TABLE 1
Prevalence of Breastfeeding in England and Wales 1975, 1980 and 1985[6] (%)

	1975	1980	1985
Birth	51	67	65
1 week	42	58	57
2 weeks	35	54	53
6 weeks	24	42	40
4 months	13	27	26
6 months	9	23	22
9 months		12	12
Base (number of infants)	1 544	3 755	4 671

delivery. Nevertheless, many mothers are eager for information at this stage.

Factors Influencing the Choice

Social and cultural factors appear to be major influences on the choice of whether to breastfeed or not. In Great Britain there is a steady downward gradient in breastfeeding rates from social class one to social class five, from the South-East of England to Scotland, from mothers over the age of 25 years to mothers of less than 18 years of age and from mothers with education beyond the age of 18 years to mothers with a shorter school education. Mothers of first babies are more likely to breastfeed than mothers of second or subsequent babies.[3] This tendency for mothers from upper socio-economic groups to breastfeed is consistent throughout the developed world. It is reversed in the developing world where mothers from upper socio-economic groups tend not to breastfeed.[9] In Britain, medical factors appear to play only a small part in influencing a mother's choice.

The Professional Educator's Role

Midwives, health visitors and health educationalists play a key role in education especially in helping mothers make their choice about how to feed their babies. It has been this country's consistent policy to encourage breastfeeding and so a degree of bias is quite proper.[6, 10–12] Education for mothers in this country about infant feeding should not be applied uncritically to other countries where circumstances are different. 'Breast or Bottle?' attracts strong protagonists and the educators may come under pressure to seek to influence the mothers' choices unduly.

EDUCATION ABOUT BREASTFEEDING

Mastering the skill of breastfeeding needs support, encouragement and the type of sisterly advice which many would not describe as formal education. The core of knowledge to be passed on to mothers who wish to breastfeed is difficult to define. It involves personal and attitudinal adaptations which vary for each individual. Those best placed to help mothers may be those women who have themselves breastfed and those women who have successfully helped others to breastfeed. A report in 1943 criticised much of the educational material about breastfeeding then available and some bizarre examples were quoted.[10] Nowadays, more sensitive and flexible programmes of education are being used.[13]

Who Educates the Mother How to Breastfeed?

In such a personal activity as breastfeeding, mothers are likely to be influenced most by those with whom they have the closest contacts. It has been shown that mothers often turn to family members and friends as a source of advice about infant feeding.[2] This contribution is valuable, although the advice usually needs to be supplemented with professional expertise. Teachers at school and health educators have a modest but important role. The major source of education for parents in this country is the health professionals and in particular midwives and health visitors.[14] Further help is available in a patchy distribution through voluntary groups dedicated to breastfeeding such as the Breastfeeding Promotion Group of the National Childbirth Trust, the La Leche League and the Association of Breastfeeding Mothers.[6]

The Educational Medium

The non-commercial sector

The most effective education about breastfeeding is through person to person individual support. This can be very time-consuming because, for instance, a feed may need to be watched through before the mother can be advised. Mothers who want to breastfeed are usually keen to find out all they can about how to succeed. There is a ready market for written information from full-size books,[15] to magazines and leaflets. The Health Education Council produced the *Pregnancy Book* in 1984[16] and the Scottish Health Education Group has produced the *Book of the Child*;[17] these deal with child care in general and there are sections on infant feeding. Written material needs careful targeting and, if the population is very diverse, varied leaflets on the same subject may spread the information more widely. The voluntary breastfeeding promotion groups provide material which tends toward the more educated parent. There is probably a need for more material for the less literate parent.

Radio and tape recordings, television and video record have all been tried and the artistic results can be impressive. The Open University has pioneered a module for child care through its Community Education Programme which is suitable for both educators and parents on their own. The material can be used effectively for groups of parents guided by a health professional such as a health visitor. Its original programme 'The First Years of Life' ran for ten years and it has now been replaced by a new programme, 'Living With Babies and Toddlers', to meet the steady demand.[18]

The commercial sector

There is a large output of educational material about breastfeeding financed by the manufacturers of infant formula milks. Commercial interests have a long-standing record of support for humanitarian and educational projects in the health field. However, in the case of infant feeding, education about breastfeeding would appear to be in conflict with the manufacturers' commercial interest to extend the market for breast-milk substitutes. The industry comes under pressure to support the principles of the World Health Organization's 'International Code of Marketing of Breast-Milk Substitutes'[19] which has, as its first aim, 'the provision of safe and adequate nutrition for infants, by the protection and promotion of breastfeeding'. The Code limits the marketing of breast-milk substitutes lest these activities have the undesirable effect of influencing mothers not to breastfeed.

EDUCATION ABOUT ALTERNATIVES TO BREASTFEEDING

Infant Formulas

The only foods suitable for the young infant are breast-milk or manufactured products, known as 'infant formulas' which have been specially formulated to provide a sole source of nutrition for infants. Compositionally these take breast-milk as their model. In developing an appropriate formula the nutrition scientist, health professional and food manufacturer must all co-operate. Human milk is characterised and then the manufacturer prepares what is feasible within the constraints of matching a biological substance. Finally the health professional assesses the product as a sole source of nutrition for infants. The usual base for infant formulas is cows' milk. In the 1970s the manufacturers of infant formulas modified their products to bring them closer to the composition of breast-milk. Most infants in this country do well when fed the modern infant formulas, although these products differ in many respects from breast-milk.

Education about Infant Formula Feeding

Education about artificial feeding can be didactic; the relative merits of the available products can be defined and the preparation of an infant feed follows recipe-like steps. Most mothers seem to get it right in this country. In other countries where parents may have a lower level of education and literacy, and where the facilities are less satisfactory, infant formula can be a hazard to the infant. Inappropriate products may be given in unhygienic circumstances.[7]

Bottle-feeding mothers turn to the health professionals, the family members and friends for advice and support. Parents know that baby milk brands are often modified and health professionals and other friends with babies on infant formula, are likely to be their preferred sources of information. Chemists, as retailers, are often asked for advice. Most bottle-feeding mothers use the packaging of the infant formula as a ready source of information.[6]

The manufacturers have a responsibility to inform and educate those advising mothers. The information about artificial feeding needs updating to take account of product development and new scientific thinking. In exceptional circumstances, warning of hazards from, for instance, contaminated products, must be disseminated without delay.[20] Close links between professionals from health and education and also with nutrition scientists, manufacturers and retailers are essential.

The Education Medium

Written information in books, magazines, leaflets and on food packaging is plentiful and good. It is often prepared for free distribution by the manufacturers of infant formula in collaboration with health professionals. Short films are available from manufacturers or through the Open University and other non-commercial sources. This medium has been particularly valuable for groups where there are literacy difficulties or in ethnic minority groups whose first language is not English.[21]

WEANING AND THE TRANSITION TO A MIXED DIET

Changes in the Diet and in the Infant

Young things delight us by growing and changing. At the same time, the baby shows an increasing authority; if an infant spits out its first taste of puréed carrots, is it because it is not yet ready for solids or is it because the child does not like carrots? The child is changing from a passive receptacle of good care to an increasingly self-determining individual. Clearly, the degree of self-determinism in a four-month-old baby and a fourteen-year-old adolescent is very different but parents need confidence in themselves and in their child to allow the process to develop.

Initially, parents want to know when to introduce foods other than breast-milk or infant formula and how rapidly to diversify their child's diet. By the end of the first year most infants will be pushing everything that they can find into their mouths and there need be no worries about diversifying their diets. A parent's concern now becomes what the child should *not* be eating. For instance, if they are worried about additives, they can be reassured that manufacturers have considered it prudent to limit the number of additives in infant foods and that all have been independently reviewed for safety.[22] Some infant diets are intentionally restricted for cultural or medical reasons. Those responsible for these infants may need advice about how to ensure a nutritionally adequate diet especially if the preferred foods are not readily available in the shops. These advisers will themselves need education about their client's beliefs and needs.[23, 24]

Nutrition and Health

Physical and mental development

Education is more readily received when the purpose of the advice is clear and the explanations are credible. In the first days of parenthood parents are very receptive, many regard their child's continued survival to be a sufficient goal. In the later stages of infancy and during childhood the objectives are physical and mental growth and maturity. Parents are invited to bring their babies to child health clinics where their physical and mental progress can be reviewed. These clinics provide good opportunities for nutrition advice. Although experts are themselves unsure about the significance of height and weight in relation to health, a workable goal for an infant diet would be that it maintains an infant's growth and development in line with that of its peer group.

Nutrition and illhealth

Long term objectives in relation to an infant's diet are difficult to define now that major nutrient deficiencies are rare. The Medical Research Council of Great Britain has examined the relationship between infant nutrition and cardiovascular disease in adult life but this work is at an early stage.[25] There is presently intense interest in the relationship between nutrition and health but those advising parents should be convinced of benefit before recommending changes in the traditional weaning diet.[26] The toddler is vulnerable and growing rapidly.

Eating habits

If infants and children are programmed in their eating habits from an early age, young childhood may be the best time to influence future patterns of behaviour. Until more is known about these questions it is difficult to define a clear educational role but this uncertainty does not inhibit a high degree of public interest in these matters.

SOURCES OF EDUCATION ABOUT INFANT FEEDING IN THE UNITED KINGDOM

The infant, its parents, society at large, and the educators themselves should each be discrete educational targets.

Educating the Infant

The scope for nutrition education of the infant may, at first sight, appear limited although infants learn faster than ever in later years and most can quickly be taught patterns of behaviour. For instance, until the 1980s infants were fed at set intervals and for set times; most babies quickly learnt these routines although it now seems likely that they may have exerted a deleterious effect on successful lactation. Breast-milk is produced in response to suckling and where this has not been frequent enough, the supply of milk diminishes. It could be suggested that, had the infants been less ready learners and had they demanded feeding more frequently, breastfeeding might have been more successful. (The commonest cause of giving up breastfeeding is the mother's perception of inadequate milk.[3]) In later infancy it has been suggested that permanent eating habits begin to be established including the frequency of eating and the readiness to accept a varied diet. Other parents are keen to avoid or limit sweets because of the damaging effect on teeth, but the extent to which the avoidance of sugar becomes a learned habit, which is carried through to the older years of childhood, has yet to be determined.

Educating the Child's Parents

Choice of feeding method

Infant feeding is a suitable subject for interactive education because it is influenced by attitudinal factors, and because there is scope for a personal interpretation of child care.[27] Infant feeding needs to justify its inclusion in the school curriculum: it provides only limited benefit for the individuals being educated but there are long-term benefits for the next generation of children. Education later on may be too late to be effective.

How to feed a baby

Pregnancy is a good time for education about the chosen method of feeding. Parents are eager to learn and there is much written, verbal and visual recorded material. As well as government sponsored materials such as the *Pregnancy Book*[16] there are locally prepared leaflets and other materials prepared by professional bodies.[28] Voluntary groups provide leaflets and arrange antenatal classes. The manufacturers of infant formulas and other baby products provide educational resources which are popular with parents. The presentation is usually more lavish with colour, photographs and large format; at the same time a proportion of the publication is devoted to advertisements which some mothers find helpful and others consider an irritation.[29] The readability of baby books for parents has been commended in comparison with other health-care pamphlets[30] and the clear text should help those who may be in greatest need of education. Notwithstanding these varied resources targeted on parents, including those with only modest educational backgrounds, the health services in this country place the main thrust in education about infant feeding in person to person contact through the health professionals.

Educating Society

Child care has traditionally been a responsibility delegated to mothers. Often that care is provided out of sight of the general society. This country has been no exception especially during the 19th century: 'Children should be seen and not heard'. More recently, mothers have enjoyed greater opportunities for social contacts and for gainful employment outside the home. At the same time few mothers have shed their responsibilities to nurture and care for children. If a mother is to move freely in society she will need to take her baby with her and she will need public facilities. Women have reported that, even in shops where the majority of shoppers are women, infant feeding in public, especially breastfeeding, can be viewed with disfavour. Boots and Mothercare have been among the leaders in providing accommodation for mothers to care for their babies. This has not always been welcomed whole-heartedly by the most eager breastfeeding proponents who claim that society should be prepared to accept breastfeeding in public without the need for special private accommodation. Where the issues contain a strong attitudinal component coupled with emotion generated by a young baby, it is not surprising that the writers of films, soap serials and magazine articles have been drawn into the arena of infant feeding. Society seems yet to be undecided about whether 'Breast is Best or is Bestial'. It is open to speculation whether education about infant feeding has any influence on this choice.

REFERENCES

1. Helsing, E. and Savage King, F. (1982) *Breast-feeding in practice,* Oxford Medical Publications, Oxford.
2. Martin, J. (1975) *Infant feeding 1975: attitudes and practice in England and Wales,* Office of Population, Censuses and Surveys, Her Majesty's Stationery Office, London.
3. Martin, J. and White, A. (1988) *Infant feeding: 1985,* Office of Population, Censuses and Surveys, Her Majesty's Stationery Office, London.
4. Hally, M. R., Bond, J., Brown, E., Crawley, J., Gregson, B. A., Philips, P. *et al.* (1981) *A study of infant feeding: factors influencing choice of method,* University of Newcastle-upon-Tyne, Health Care Research Unit; No. 21.
5. Valman, H. B. (1984) *The first year of life,* British Medical Association, London.
6. Department of Health and Social Security (1988) *Present day practice in infant feeding: third report,* Reports on health and social subjects; No. 32, Her Majesty's Stationery Office, London.
7. Grant, J. P. (1987) *The state of the world's children 1987,* United Nations Children's Fund, Oxford University Press, Oxford.
8. Martin, J. and Monk, J. (1980) *Infant feeding 1980,* Office of Population, Censuses and Surveys, London.
9. World Health Organization (1981) *Contemporary patterns of breast-feeding,* World Health Organization, Geneva.
10. Ministry of Health (1943) *The breast feeding of infants,* Reports on public health and medical subjects; No. 91, Her Majesty's Stationery Office, London.
11. Department of Health and Social Security (1974) *Present-day practice in infant feeding,* Reports on health and social subjects; No. 9, Her Majesty's Stationery Office, London.

12. Department of Health and Social Security (1980) *Present day practice in infant feeding: 1980*, Reports on health and social subjects; No. 20, Her Majesty's Stationery Office, London.
13. Royal College of Midwives (1988) *Successful breast feeding — A practical guide for midwives and others supporting breastfeeding mothers,* Royal College of Midwives, London.
14. Department of Health and Social Security (1988) *Public Health in England,* Her Majesty's Stationery Office, London.
15. Llewellyn-Jones, D. (1983) *Breastfeeding — how to succeed,* Faber and Faber, London.
16. Health Education Council (1984) *Pregnancy Book,* Health Education Council, London.
17. Scottish Health Education Unit (1980) *The Book of the Child,* Scottish Health Education Unit, Edinburgh.
18. Open University (1987) *Living with babies and toddlers,* Open University, Milton Keynes, United Kingdom.
19. World Health Organization (1981) *International Code of Marketing of breast-milk substitutes,* World Health Organization, Geneva.
20. Rowe, B., Begg, N. T., Hutchinson, D. N., Dawkins, H. C., Gilbert, R. J., Jacob, M., *et al.* (1987) *Lancet,* **ii**, 900–3.
21. Bahl, V. (1987) *Asian mother and baby campaign,* Department of Health and Social Security, London.
22. Ministry of Agriculture, Fisheries and Food (1981) *Food Standards Committee Report on Infant Formulae,* Her Majesty's Stationery Office, London.
23. Henley, A. (1981) *Asians in Britain; Foods and diets,* Department of Health and Social Security: King Edward's Fund for London, National Extension College, Cambridge.
24. Francis, D. E. M. (1986) *Nutrition for children,* Blackwell Scientific Publications, Oxford.
25. Medical Research Council Environmental Epidemiology Unit (1987) *Infant nutrition and cardiovascular disease,* MRC Environmental Epidemiology Unit, Southampton.
26. Clarke, B. and Cockburn, F. (1988) *Nursing Times,* **84**, 59–64.
27. Finch, J., Clayton, S. and Clements, J. (1981) *Health Education Journal,* **40**, 19–23.
28. Royal College of Midwives (1987) *Your first baby,* The Newbourne Group, London.
29. Rodway, A. and Vaizey, M. J. (1987) *Baby's first year,* Bounty Publications Ltd, Diss, Norfolk.
30. Nicoll, A. and Harrison, C. (1984) *Devel. Med. Child. Neurol.,* **26**, 596–600.

Report of Discussion

Rapporteurs: C. A. ZAROR[a] and J. G. W. JONES[b]

[a]Department of Food Science and Technology, [b]Department of Agriculture,
University of Reading, UK

The papers presented were commended by participants in the discussion for their exposition of the place of formal education in relation to the operation of the food chain. The discussion focussed in large measure on the post-farmgate sectors of the food chain.

The balance between the theoretical and vocational aspects of the food chain was regarded as important. Whilst teaching currently needed skills ensures the short-term marketability of students, the theoretical foundations may turn out to be inadequate in the longer term. The vocational content of courses is also important because of the considerable effect which it has in motivating students and attracting them to the courses in the first place. To find the correct balance between the conceptual, theoretical elements and the vocational ones was a task that should be carried out in close collaboration with industry.

The problem of motivating and recruiting students was seen as requiring much closer contact between schools and the tertiary education institutions. Information on the range and nature of courses on offer should be made readily available in a variety of ways and the assistance of the industry should be sought in doing so. In courses in agriculture an acquaintance with some of the practical aspects of the industry had usually been made before the courses started.

At the outset of their professional lives, graduates found themselves dealing with very specific tasks, particularly in the food industry. It was not possible to cover the spectrum of such tasks in the relatively short duration of an undergraduate course. There was, too, the problem of maintaining the student's interest and here it was felt that a wide selection of subjects should be offered so as to enable students to

173

complement the core curriculum with subjects of their choice which engaged their interest.

It was recognised that graduates entering industries associated with the food chain, particularly post-farmgate industries, experienced difficulty in accommodating to the harsh realities of industrial life, in particular, the need to work in teams, under pressure with strict deadlines and discipline. However good the educational background of the new graduate, real work experience cannot be acquired in the university or polytechnic environment. Industrial training during their undergraduate education, as in the case of sandwich courses, helps students not only to gain that experience but also to identify at an early stage in their courses those areas of training that may need to be strengthened after graduation.

Post-experience education came in for considerable discussion. The technological changes taking place in industry make it necessary for education to be regarded as a continuing process throughout the career of an employee. Whilst the earlier stages of formal education are characterised by a necessarily broad approach, it was argued that post-experience education should provide for the acquisition of more specific skills and for updating skills already acquired. Two major difficulties were identified in providing post-experience education; first, post-experience education may compare unfavourably with the in-factory training provided by employers; second, there is a perception that opting out of industry to pursue post-experience education may be detrimental to career prospects.

Industry has to be encouraged to work more closely with further and higher education in (1) teaching and (2) providing specific guidelines for course design. Final decisions on course contents have to be made by the teaching institution both for academic reasons and for reasons of resource provision. Many examples exist of professionals from the food and farming industries being directly involved in teaching as invited lecturers, in seminars and workshops; such contacts enable direct communication between the industries and the teaching establishments which, in the context of course development, should provide a fertile ground for discussion and exchange of ideas. Unfortunately, the current picture is one of industry being reluctant to participate to the extent which is required.

The task is formidable: to strengthen communication at all levels in the food chain and, in particular, the links between schools, further and higher education, industry and the service agencies participating in the food chain.

THEME 3

Implications for Technology:
Priorities for R & D

Introduction

H. E. NURSTEN

Department of Food Science and Technology, University of Reading, UK

The Food Chain that is the subject of this Conference is the anthropocentric one, stretching in various ways from farm via the markets, the food industry and the retailers to the consumer. Importing, exporting, fishing, hunting, ranching, gardens and allotments, and catering all play a role in it. The Food Chain is not independent. It is embedded in the environment and it is tethered to other human activities, such as the production of alternative crops, of wool, and of hides and skins.

The Food Chain is far too complex to be dealt with in sufficient detail in just one conference, so there is a great need to be selective, but nevertheless retaining the complexity at the back of one's mind.

Clearly the primary objective intrinsic to the Food Chain is to feed the population. Food security is paramount. It follows that one of the most important aims must be to have integrated management of the Food Chain, that is, to abolish so far as is feasible the traditional barriers both at the farm gate, between the processor and the retailer, and between the retailer and the consumer.

Since we are now largely in an era of plenty, the emphasis is no longer on simply feeding the population, but quality is on the ascendant, as far as the consumer is concerned. Food quality can be considered from many points of view, six in particular:

Nutritional	Biological
Sensory	Chemical
Microbiological	Physical

Foodstuffs (including drinks) are traditionally sub-divided into eight commodity areas:

Meat, poultry, fish
Dairy products, eggs
Cereals
Fruit, vegetables
Oils, fats
Sugars, preserves, confectionery
Beverages, soft drinks
Alcoholic drinks

Each combination of foodstuff and food quality needs to be explored in terms of its Food Chain, particularly from two points of view, flexibility in relation to the possibilities for new products and pressure for change from and on the individual links in the chain.

Clearly, the field to be covered is too vast to be encompassed within our programme. We have therefore restricted ourselves to parts of the first five of the eight commodity areas and to highlight for each the principal problems and to give some indication of how they might be solved.

This is done against a background of several important reports that have been published recently.[1-4]

REFERENCES

1.   MAFF (1985) Food Safety Research Consultative Committee (the Norris Committee), Report to the Priorities Board, 42 pp.
2.   MAFF (1986) *Final Report of the Food Composition and Processing Research Consultative Committee (the Gorsuch Committee),* 50 pp.
3.   Priorities Board for Research and Development in Agriculture and Food (1987) *Second Report to the Agriculture Ministers and the Chairman of the Agriculture and Food Research Council,* 50 pp.
4.   *Research for the Food and Drink Industry* (1987) A Food and Drink Federation submission to the Priorities Board for Research and Development in Agriculture and Food and to the Ministry of Agriculture, Fisheries and Food (the Righelato Report), 25 pp.

Fruits and Vegetables

V. D. ARTHEY

Campden Food and Drink Research Association, Chipping Campden, UK

ABSTRACT

This paper looks at the ways in which the presentation of fruits and vegetables to the consumer are changing. The increasing range of styles in which some vegetables can be sold is noted, but of particular interest is the growing demand for quality. Visual quality can now be ensured but there is still room for improvements in sensory quality and manufacturers and retailers are using chilled products, sometimes packed in modified atmospheres, as one way of achieving what is perceived as freshness. This is all bound up with nutritional value and healthy eating and thus organically produced fruits and vegetables are another means of attracting consumer interest.

Modern horticultural techniques require not only traditional skills but an increasing knowledge of science and technology and this is also true for the marketing of todays products in different forms with different packaging materials. The skills of yesterday are being blended with the science of tomorrow.

INTRODUCTION

It has been claimed that approximately 75% of all UK agricultural produce is processed in some form or another and an estimate of the production of fruits and vegetables in different forms is presented in Table 1. If it is accepted, however, that fresh fruits and vegetables are 'processed' when they are wrapped in polythene or paper, polished or waxed before sale, then this figure could well be much higher and the very nature of the many 'processes' given to edible horticultural

179

TABLE 1
Production of UK Fruits and Vegetables in Different Forms in 1986

	Fruit	*Vegetables*
Fresh	464 700 tonnes	10 150 000 tonnes (including potatoes)
Canned	34 756 tonnes (net can contents)	781 562 tonnes (net can contents)
Frozen	19 912 tons[a] (including fruit products)	508 767 tons (1985 data)
Dried	—	29 810 tons (dried)
Jam	99 645 tons	—
Pickles	—	83 107 tons

[a]One ton = 1.016 05 tonne.

products suggests the variety of ways in which these products can be offered to the consumer. Today cabbage can be sold as whole heads of different types (e.g. red, pointed, Dutch white, savoy and others), half heads, shredded, in modified atmosphere packs or as coleslaw (chilled), frozen as shredded pieces either in block or free-flowing style, and canned or fresh as sauerkraut. The diversity of packs of fruits and vegetables is increasing year by year. Together with this multitude of products comes a very obvious demand for improved quality and a consumer willingness to pay for it.

THE DEMAND FOR QUALITY

In terms of quality demand for fruits and vegetables, it was probably the canning industry which first set the ball rolling in the 1920s. The canning industry was founded on the use of surplus raw material from the fresh market, but it soon became apparent, that, if a reasonable quality product was to be offered to the consumer, then some control would be necessary over the raw material to be used in the processing operation. The canners began to specify what they really wanted and, as growers sought to meet that specification, so the quality of the canned product improved. The time came when the original situation was reversed and the fresh market had to accept the unwanted portion of the canning crop. This remained true for many years after the Second World War, but, with the advent of the freezing industry and more recently the emergence of a controlled fresh produce market, each outlet began to specify its precise requirements and today we see the specific needs for the canning, freezing, chilled and fresh market outlets all being met by the agricultural/horticultural industry. In these well ordered markets — and they are well ordered in the UK when compared with many other countries — there is a smooth order of business transaction. The supermarket determines the future consumer demand in consultation with its sales force. This demand is translated into production requirements and contracts are drawn up with growers capable of meeting not only the quantities required but the specific quality characteristics of each fruit and vegetable demanded by the specification supplied to him. Making the choice of what variety to grow is probably the most important raw material decision to be made in relation to final product quality, and it is aided by the Technical Manuals which give in detail the results of work undertaken by the Campden Food and Drink Research Association on the suitability of varieties for the canning and freezing industries. Larger market outlets — retail supermarkets and large processors — will monitor the crops to ensure that all is being done to maximise the quality required. Harvesting, handling and storage, where necessary, are all scientifically controlled and processors check quality of delivered loads at factory gates where they are rejected if not up to the standard required. Preparation and processing is governed by good manufacturing practices and the product is presented to the consumer in one or more of a wide range of packaging materials. This chain of operations from initial grower to consumer purchase represents a smooth progression through its many links, although pressures for change do occur from time to time to test the tenacity of the individual links. Whilst it would take a market researcher to identify precisely the origin of each pressure for change; that is whether it comes from the government, the grower, the retailer or the consumer, it is suggested that they probably originate in the main from the retailer. Of course, there are exceptions to this and these will be referred to later.

Many years ago, a weak link in the chain was the barrier at the farm gate. This has been largely overcome by the realisation that both processor and grower are part of a single food production line and an understanding of each other's problems has been essential. Today the pressures at this point are rarely technical; the negotiations which still occur annually are over the contract prices for major fruits and vegetables for the processing industries. Sometimes heated discussions are necessary to arrive at an equitable price acceptable to both sides.

More topical are the pressures for specific quality requirements in an increasingly sophisticated food market. The retail outlets are primarily responsible for this demand for improved quality, although some of it originates from the consumer. There is no doubt that quality is in demand; several retailers are renowned for it and others are clearly seeking to emulate the best. In fresh fruits and vegetables visible quality is often as good as it can be and, where it fails, the reasons are known and generally due to lack of control in the growing, harvesting, handling or packaging/processing. Improved quality, therefore, can only come effectively from an improvement in the 'hidden' qualities of internal colour, flavour, texture and other eating characteristics. There are pressures for change in this area where research needs to be undertaken to determine the precise factors which lead to improvements in sensory qualities, particularly flavour.

METHODS OF PRESENTATION

Today's processed fruits and vegetables do not come solely in cans or frozen packs. They come in so-called fresh forms, trimmed, wrapped, in trays and in the future they may have modified atmospheres to lengthen the product shelf life. They also come in chilled forms where temperature control is another way of extending the shelf life. We may see irradiated products in the future, when the method is approved by the UK government as it is in other countries and provided the industry can overcome the barriers erected by lobbyists.

CHILLING

The growth of the chilled food market has been particularly spectacular in recent years, with much space being devoted to this sector in retail supermarket outlets.

A chilled food is defined as a perishable product intended for distribution and sale by retail or through catering outlets at controlled temperatures in a range selected between $-1°C$ and $+10°C$ and frequently $0°C$ to $+8°C$. It is now estimated that nearly 50% of all household expenditure on food is on chilled foods, and fruit and vegetables often form a major part of these products. For example, there is a growing demand for coleslaws, fruit yoghurts, vegetables in mayonnaise and other dishes, which give fruits and vegetables a ready-to-eat image coupled with recipe innovation. However, there are special manufacturing requirements for these foods which take into account: the perishability of the raw materials; the necessity to maximise the sensory quality by minimising handling and processing procedures; the potential for spoilage and/or food poisoning; the necessity to dispatch the finished products without delay; and the special cool chain requirements. For example, if the pH of a newly produced ready-to-eat recipe salad becomes too high, then microbiological hazards can arise, which could result in food poisoning outbreaks. Similarly, abuse of the handling and processing procedures could impair the efforts to retain crisp texture, brightness and that unique and elusive fresh flavour.

MODIFIED ATMOSPHERES

Much work is currently being devoted to the development of the use of modified atmospheres as a technique which together with chilling will extend the shelf life of fresh fruits and vegetables. The gases under investigation are oxygen, carbon dioxide and nitrogen, but the method has not yet become fully accepted by the food industry for fruits and vegetables (as it has for meat and fish), because of the cost of the equipment and its installation in the packhouse, even though the method can double the expected shelf life of the product.

THE COOL CHAIN

An essential part of handling of both chilled and modified atmosphere packaged products is the use of an efficient cool chain. The cool chain is an organised system for the distribution of chilled foods at controlled temperatures, from the point of production to retail sale or use in catering, and consists of initial cooling, cool storage, refrigerated transport and warehousing, and retail display on refrigerated counters with precautions to avoid or minimise exposure to ambient temperature during transfers. Unfortunately, there are many opportunities for the abuse of the cool chain and these will impair not only the quality of the product but even the image of a brand. It is also interesting to note that sometimes, where the product has been chilled satisfactorily and is conveyed without abuse, if it is then to be displayed at ambient on the supermarket shelf, its quality may deteriorate quite rapidly.

Chilled and modified atmosphere foods carry with them their own particular problems and difficulties and often the more sophisticated the product, the greater the problems. They have a two-fold purpose. One is to extend the time over which the product can be offered to the consumer, and the other is to retain for as long as possible the initial quality of the fruit or vegetable. To use these sophisticated systems for inferior quality produce is not cost effective and only the best is, or should be, subjected to such treatment, where the price commanded is worth the handling on-cost.

PACKAGING

Most fruits and vegetables are sold today in packaged form and the type of packaging used is an integral part of their manufacture. Chilled fruits and vegetables are frequently sold in overwrapped trays and fresh vegetables are often marketed in polybags. Frozen fruits and vegetables are also sold in polybags and there is a considered opinion that quality could be retained to a greater degree if the polybag were of an improved grade, or even foil laminated. There has been a remarkable development in packaging in the last few years with high-barrier plastics being introduced for heat-processed products and today it is possible to purchase shelf-stable fruit products in clear high-barrier plastic containers the ends of which can be seamed on conventional canning lines. With modified atmosphere products, it can often be quite difficult to find packaging films with the required permeability for gaseous exchange shown by research to be necessary to give an extended shelf life to the product. Packaging research is expected to continue unabated for well into the future to ensure protection, safety and quality of the foods we eat.

ORGANIC AND ASSOCIATED FOODS

The current surge of interest in organic foods is probably more closely related to consumer demand engendered by lobbyists. Nevertheless it is anticipated that the market will reach at least 1% of the fresh market, which is not insignificant in monetary terms. Retailers are attempting to meet this pressure for change by supplying what is required, but they find difficulty in securing supplies of truly organically produced products. The vogue is becoming established, but only time will tell whether it will become permanent. Will the consumer really believe that quality in a lettuce is indicated by the presence of a slug or some aphids? She may accept this situation in produce from her garden, but it is doubtful whether she will willingly accept it from her retail outlet.

Closely associated with organically grown foods is the demand for reduced amounts or absence of additives and contaminants. This pressure comes from the lobbyists, but is being vigorously reinforced by the retailers. In fresh fruits and vegetables it amounts to a demand for the absence of pesticide and fertiliser residues. Although the consumption of organic foods will largely meet this demand, it has wider connotations. Consumers of conventionally produced foods are seeking a reduction in these residues and in future this will lead growers to reduce seriously the number and concentration of applications to their crops. Such pressures together with those for organic foods may well lead to lower yields, greater trimming and inspection and therefore higher costs, so that the pressures of change may be not only on the grower, but could revert back to the consumer, who will have to decide whether he/she is willing to meet the increased outlay.

The environment in which some crops are grown has received the attention of horticultural researchers and has been successful in that innovative methods of crop production have been introduced commercially. The production of tomatoes in a carbon dioxide enriched atmosphere is not as new as the growing of the same crop in soil-less culture, such as hydroponics. In this way, it is possible to control the growth of the plant to ensure maximum yield and recently work has been done to show how the components of the solution and their concentration can affect both yield and quality. By carefully modifying the solution, the perceived qualities of the fruits can be noticeably enhanced or reduced. In addition, some pest and disease problems can be controlled by further additions to the growing solution. It can, of course, mean that only sufficient salts or compounds are used to effect the healthy growth of the plants and that surpluses of such costly compounds are avoided.

NUTRITIONAL VALUE

The demand for absence of additives and contaminants is matched by a positive desire to eat foods of high nutritional value. The JACNE report of 1984 recommended an increase in the consumption of fruits and vegetables for a healthier heart and the 'F-Plan Diet', which indicated the dietary fibre value of a portion of baked beans, resulted in an increase in the sales of canned beans. Fruits and vegetables are good sources of fibre, vitamins and other ingredients of high nutritional value and their consumption is a positive factor in healthy eating. It is important, however, to have an understanding of the effects of the way in which the product is offered to the consumer on the nutritional or compositional value of the food. Is the identical vegetable or fruit more nutritious when canned, frozen or fresh? Work is currently on-going to determine the relative values of such products when consumed in different forms and this information will eventually find its way on to nutritional labels. As the consumer becomes even more health conscious, nutritional value is going to become an even more important ingredient of product quality.

Pressures for improved general quality, for organically produced foods, foods without additives and contaminants and with high nutritional value could be met in the future by the introduction of new varieties which have been genetically engineered. Already it is possible to produce plants which are resistant to certain herbicides, for example, resistance to glyphosate has already been achieved in the tomato and an American company expects to commercialise glyphosate-tolerant crops in the early 1990s. It can only be a matter of time before we shall see new varieties exhibiting more closely the precise requirements of the market for which they are destined. They will also have increased — perhaps total — resistance to pests and diseases and be able to produce economic yields without artificial fertiliser.

OTHER FRUITS

Whilst it is unlikely that exotic fruits and vegetables — currently receiving considerable attention — will be produced in the UK, unless grown in a protected environment, it would be comforting to believe it possible to improve production techniques to enable the UK industry to compete more effectively against imported temperate crops. The economics of production behind the iron curtain, where such crops are subsidised to obtain western currency, however, may be difficult to overcome.

GENERAL

There will be some pressure to change as the horticultural industry is increasingly burdened to fund its own research. ADAS has begun to charge for many of its services and advice may no longer come free. The Horticultural Development Council also takes a levy to support research and development in husbandry and other aspects of direct importance to the production of fruit and vegetable crops. At one time, the grower chiefly employed his skill to be successful, but, today, his business is increasingly dependent on science and technology. In the future, he will have to employ the right proportions of skill and science, not only to compete with overseas suppliers, particularly from southern Europe, but also to fund the research which will take his industry successfully into the twenty-first century.

Dairy Products

FRANK HARDING

Milk Marketing Board, Thames Ditton, Surrey, UK

ABSTRACT

The dairy industry is the largest single sector of the UK food and drinks industry with milk from UK farms being valued at over £2 billion, and the retail value of products made from it being £4· 5 billion per annum. The dairy industry has had to adapt to dramatic changes in recent years, the most significant of which has been the imposition of quotas restricting the amount of milk produced within Europe. This has switched the emphasis of research from utilisation of a surplus product to a focussing of interest on the reduction of the cost of milk production and usage of milk in high-added value outlets. The industry has invested heavily in improving the quality of the raw material, researching the health benefits of its products, and the development of new products in the light of changing market needs. The industry has set clear strategic objectives for research and development and relies heavily on private sector investment to provide underpinning science to sustain the development work needed to maintain the UK's competitive position in the market place. There are many examples of the fruit of private sector investment being taken up by the industry and a case is made for continued future investment in support of an industry which is a major part of the UK's food chain.

THE SIZE AND STRUCTURE OF THE DAIRY INDUSTRY

The dairy industry is the largest single sector of the UK Food and Drinks Industry. It is a low margin industry and competition from dairy industries in the EEC countries, many of which enjoy increasing

189

government support to their technology base, has put additional pressure on the UK dairy industry in recent years.

Dairy products account for about 16% of the UK consumer expenditure on food and additionally a large proportion of the beef consumed in the UK is derived from the dairy herd. The industry is a large employer in production, processing and manufacture of milk and milk products and there are several support industries, e.g. animal feeding stuffs, vehicle and machinery manufacturers, glass and other forms of packaging. The total number employed exceeds 200 000.

The United Kingdom is a net importer of dairy products, but the value of the milk produced by UK farmers is over £2 billion and the retail value of the products made from it is £4·5 billion per annum. The traditional outlets are liquid milk, butter and cheese, which account for around 90% of the milk produced with a small but increasing contribution to the industry margins coming from high added-value products. It is in this latter area where considerable expenditure is made on new product development. Exports of dairy products are relatively small, but have grown in the last few years with the help of R & D.

Dairy products make a major contribution to the energy, protein, vitamin and mineral content of the UK diet: 56% of dietary calcium is supplied by dairy products. Being a highly perishable raw material produced every day of the year, demands on processing and manufacture are high in terms of product quality, product safety and process control.

MILK PRODUCTION

The dairy industry has had to adapt to dramatic changes over the last twenty-five or more years (Table 1). Spurred on by calls for food from our own resources, the dairy industry strove for increased output, trebling the size of the average dairy herd and increasing the yield per cow by over 40%. This resulted in an increase in total milk production in spite of a dramatic reduction in the number of dairy herds. The gradual increase in milk production of about 2% per year was slowly moving the UK towards self-sufficiency, as we have always been net importers of dairy products. However, increased milk production within the EEC, even by countries who were net exporters, led to growing surpluses, which had to be met by increasing consumption or decreasing production in order to bring the two into balance.

In 1984 the decision was taken to restrict production over all EEC

TABLE 1
Dairy Farming (England and Wales) 1959/60 Compared with 1984/85

	1959/60	*1984/85*
Average herd size	21	67
Dairy farmers	105 576	37 815
Number of dairy cows	2·59 million	2·6 million
Yield per cow	3·12 litres	4·85 litres
Total output	8·07 million litres	12·605 million litres

countries by imposition of quotas. UK milk production was cut by 7·5% with a further 9·5% cut being implemented. Quotas hit the UK dairy industry particularly hard, since we have always produced less milk than we consume. It is estimated that at the end of this current round of quota cuts, the UK will only produce 85% of its needs, whereas Holland, for example, produces over 300% of its domestic requirement. Currently we import about 25% of our dairy product consumption, hence some of our production goes to intervention storage. The planned expansion of production to meet domestic requirements has therefore been abruptly ended. The industry has changed from being production led to being market driven.

Clearly, this has changed the objectives of milk producers, who, with a limit set on the volume of production, can only improve profitability by improved efficiency and cost reduction. The emphasis clearly must be on reducing the cost of milk production. However, there are demands from the market place for high product quality and extended shelf life, hence raw milk quality parameters aimed to meet these changing market trends actually increase the cost of milk production.

There are ways in which the dairy industry can improve its products but these may be in conflict with reducing costs. The spreadability of butter can be improved by modifying the feed of the cow in such a way as to increase the monounsaturated fatty acids in milk at the expense of saturated fatty acids. This increases cost; however, in such a case, it may be argued that any cost increase for such new products should be met from a higher market price.

Centralised testing of milk was introduced by the Milk Marketing Board in 1982, since when the bacterial count of the National Milk Supply has been reduced five-fold (from 94 000 to 17 000 organisms/ml). Similarly, trace levels of antibiotics in milk have been reduced five-fold.

Such quality payment schemes cost dairy farmers over £2·5 million per year plus the increased costs of detergents, replacement liners, rubbers, etc. In this case, the extra cost to the dairy farmer has to be regained by improved efficiency.

Much of the improved efficiency in past years has been due to improved yield per cow by virtue of genetic selection. Major strides in genetic engineering now enable scientists to accelerate such improvements even faster. The milk production stimulant, bovine somatotropin (BST), a naturally occurring protein hormone secreted by the pituitary gland of the cow, can now be produced by bacteria through recombinant DNA technology. BST produced by genetic engineering has been shown to raise the yield per cow by about 20%. With a quota system, clearly such a technique has economic potential to improve feed efficiency of the cow, balance out seasonal peaks and troughs in production and enable producers to adjust production more readily to meet quota. However, there are risks. The milk processor must be assured that BST-produced milk does not adversely affect products (heat stability, cheese yield, etc.). The consumer must also be convinced that BST-produced milk is not different from 'normal' milk, which she sees as 'natural' and 'healthy'.

Production research must, in general, under a quota regime be aimed at reducing the cost of milk production. However, in using the tools of science to achieve this end, we must have an eye on the consumer and the markets for the product. The closure of the National Institute for Research in Dairying and formation of the Institute of Food Research and the Institute for Grassland and Animal Production separated milk production and milk utilisation research south of the border, whereas they still remain linked at the Hannah Research Institute, Scotland. For the reasons given earlier, there is a need to maintain close contact between production and utilisation research.

MARKET CHANGES

There are well recognised trends in the marketplace — more out-of-home eating, decline in family eating with more convenience foods, coupled with new food processing developments, more leisure time and an older average age of consumers, more foreign travel, and greater tendency for larger, less frequent shopping, hence a longer-life requirement on products. There is also a growing interest in food quality, in 'naturalness', 'freshness' and 'healthy eating'.

TABLE 2
Contribution of Dairy Products to the UK Diet (%)

	Liquid	Other milks and cream	Cheese	Butter	Total dairy products
Energy (kcal)	9·6	1·6	2·9	4·6	18·7
Protein	14·6	1·8	5·8	—	22·2
Fat	11·8	1·7	5·1	10·7	29·3
Carbohydrate	5·7	1·3	—	—	7
Calcium	36·8	5·2	13·7	0·2	55·9

Source: Ministry of Agriculture, Fisheries and Food, 1986.

Dairy products make a significant contribution to the UK diet, as shown in Table 2. They are consumed in their own right, as ingredients to add flavour (e.g. cheese on pizzas), as complements to make other foods more palatable (e.g. butter on bread, milk on cereals) and as food ingredients for their functional properties (e.g. milk in cakes and bread).

The diet/health debate has largely focussed on the role of dietary fat and coronary heart disease and has led to an increased demand for low-fat dairy products. Sales of semi-skimmed and skimmed milks now account for 21% of total sales, four times the 1983/84 figure. There is an increased demand for low-fat yoghurts, cheeses and other dairy products. Here there is a research need for improvement of texture and flavour of low-fat products. Perhaps surprisingly, there is also increased demand for high-fat products (Channel Island breakfast milk, high-fat cheeses, real dairy ice cream).

Clearly the consumer is confused by the nutritional messages she receives and there is a need for more accurate information about the role of diet in coronary heart disease compared with other risk factors. There is also a need to ensure that dietary advice is balanced in order to avoid dramatic changes which might cause other health concerns. The total removal of dairy products from the diet in order to meet dietary targets on fat would dramatically reduce calcium intake, for example, and this could have serious effects on bone health.

Changes in consumer buying matters have been demonstrated in increased sales of liquid milk in shops, which now accounts for over 20% of total milk sold, the remaining 80% being delivered to the doorstep. This has led to different and generally stricter quality parameters being set by the retailers. Requirements for extension of the

Frank Harding

TABLE 3
Milk Utilisation in the UK 1985/86

| | | Volume | |
		Million litres	%
Total ex-farm sales		15 287	
Liquid sales		6 901	45
Manufacture — Butter		4 439	29
	Cheese	2 533	17
	Milk powder	337	2
	Cream	535	4
	Other	542	3

shelf life of milk and dairy products — normally referred to as short shelf-life products — have become more common.

On-farm refrigeration of milk, strict control over bacterial counts and delivery in insulated containers in a short period of time to their processor have become the norm in recent years. The processing side of the industry has now been forced to look closely at its processing, packaging and distribution systems in order to meet consumer requirements for longer shelf-life products.

MILK PROCESSING

Liquid milk, butter, cream and cheese still account for most of the milk consumed in the UK, with other products such as milk-based desserts making a small but growing contribution (Table 3). Liquid milk, milk-based drinks and desserts offer greater financial returns than butter and cheese production, hence tend to have greater attention focussed on their development than butter and cheese.

RESEARCH AND DEVELOPMENT WITHIN THE DAIRY INDUSTRY

It is estimated that the dairy industry spent about £8 million on technological research and development in 1984. Product and process development facilities are operated by many of the larger dairy companies. Research is also contracted to the Research Associations.

The dairy industry, Milk Marketing Board and Dairy Trade Federation jointly fund research projects selected, managed and financed by the UK Dairy Industry Research Policy Committee (UKDIRPC), which has been operating since 1978. UKDIRPC operates through expert groups of scientists and technologists from the private and public sectors, who select projects to be sponsored with the public sector.

The dairy industry also has a Nutritional Consultative Panel, which assists in the identification of needs for research into human nutrition. A close interface with the Hannah Research Institute is maintained through its Consultative Panel, which has met twice a year since 1972. Contact is also maintained with the Institute of Food Research, Reading Laboratory.

STRATEGIC OBJECTIVES FOR RESEARCH AND DEVELOPMENT

The UK dairy industry has the following general objectives:

— To use high standards of hygiene and process control in converting a highly perishable raw material into appealing and nutritious products for the consumer and into ingredients to be used in food and other industries.
— To use efficient processing and manufacturing methods and to develop new unit processes with the objective of making traditional, new and improved products with the lowest possible associated costs.
— To develop new technologies and new markets, in part by identifying and transferring technology from the food, pharmaceutical, chemical and electronic industries.
— To expand exports and displace imports of dairy products.

NEEDS OF THE PRIVATE SECTOR FOR UNDERPINNING SCIENCE AND TECHNOLOGY

The private sector is not equipped to carry out the underpinning R & D required to meet its strategic objectives. The areas of science and technology for which the private sector looks to the public sector for support are as follows:

Physics and Engineering

The need — Improved knowledge of the physics and engineering of processing and manufacture: development of sensors, and of automatic and electronic controls for industrial applications. Improved basic knowledge of heat exchange operations, particularly with respect to fouling, enzyme destruction, product acceptability and food safety.

Examples — Many traditional processes such as cheesemaking and buttermaking have been engineered to convert them from batch to continuous processes. On-line sensors are therefore becoming increasingly important in controlling processes, reducing costs, reducing wastage and improving product quality and safety.

Fractionation, Modification and Waste Recovery

The need — Understanding of the science involved in fractionation and modification of milk and its constituents and in processes used for waste products recovery.

Examples — Butterfat competes in the marketplace with other fats and oils, many of which are imported. The development of new means of utilisation of butterfats will therefore be of significant commercial benefit. Hence there is a need, for example, for fundamental work on non-aqueous phase enzyme chemistry to open the way for the industry to make added-value products based on butterfat.

Biotechnology

The need — Basic technology including fermentation and its application to dairy manufacturing, e.g. involving genetic modification of micro-organisms and exploiting natural and modified enzymes.

Examples — Naturally occurring micro-organisms, e.g. the lactic acid bacteria, have traditionally been used for the production of fermented dairy products such as cheese and yoghurt. The characteristics of these organisms are being modified by genetic manipulation. This research promises to provide improved starters, more cost-effective processes, new preservation systems and fine chemical production from milk.

Composition, Component Utilisation and Interactions

The need — Improved knowledge of the constituents of milk, their behaviour during storage and processing, and as raw materials for food and other uses. In particular, emulsion technology, natural inhibitors and new uses for lactose and milk fat. Understanding of interactions between milk components and other food constituents during processing and in storage and their effects on colour, flavour and texture.

Examples — Milk ingredients, such as milk proteins, are widely used in food manufacture and underpinning research is required to characterise milk constituents and their interaction. This research would help to develop new markets for milk and stimulate development of a greater variety of added-value products.

Consumer Acceptability

The need — Understanding the basis for consumer acceptability of dairy products, involving the development of objective methods of sensory assessment, especially in relation to flavour and texture. In particular, improving the suitability and acceptability of products made in the UK for consumption in countries to which they may be exported.

Examples — Basic understanding of consumer acceptability and sensory evaluation is necessary to ensure the commercial success of new and existing dairy products.

Nutrition

The need — Understanding the relationship of diet with health, especially in relation to milk as a source of nutrients, and in this context improved knowledge of human physiology and nutrition.

Examples — Recent dietary trends are based on limited scientific and medical evidence and could have an influence on the long-term health of the nation. There is a need for the public sector to provide an understanding of the implications of dietary changes. Dairy products, for example, supply well over half of our dietary calcium and the effect of reduced calcium intake due to lower consumption of dairy products and the reduced bioavailability due to increased consumption of fibre needs to be assessed.

There is also a need for wider nutrition education of general practitioners and others offering advice on preventative medicine.

CONCLUSIONS

The dairy industry is a major part of the UK food industry and makes a very significant contribution to the UK diet. The industry is a major employer which with support industries accounts for over 200 000 employees.

Individual dairy companies have product and process development facilities and the MMB provides R & D support for the use of England and Wales milk. The UK dairy industry has developed a good system of contacts between public and private sector research and relies on the support of the public sector for underpinning science and technological R & D in order to maintain its competitiveness, to support product and process development and innovative development of added-value products. There are many examples of the fruits of these links in the past:

— The development of the DEFT (direct epifluorescent test) as a method of rapid bacterial counting (Institute of Food Research, IFR).
— The development of infra-red methods of measuring the composition of milk (IFR).
— Antibacterial systems in milk — nisin antibiotic discovery (IFR).
— Accelerated enzymic ripening of cheese (IFR).
— Cheddaring Tower for use in Cheddar cheese production (IFR).
— Stabilising alcoholic milk liqueurs (Hannah Research Institute, HRI).
— Rapid detection of post-pasteurisation contamination.
— Milk composition/cheese yield equations.

Clearly there is a need in the present environment of reduced investment in publicly funded research to ensure that strong links are forged in the restructured chain to ensure that the UK dairy industry's future in this rapidly changing world is secured.

Meat

G. HARRINGTON

Planning and Development, Meat and Livestock Commission, Milton Keynes, UK

ABSTRACT

Contrary to impressions given by the media, meat consumption in Britain has not fallen, but it has been static and not shown the growth experienced throughout the rest of Europe. The extended chain from Welsh hill to supermarket cabinet poses unique problems of quality control for a product under new pressures due to changing lifestyles and eating patterns.

The sector must be assisted to overcome imperfections of structure, adaptability, attitude and awareness that hinder the conversion of research into improved practice or products, and encouraged to co-operate in the planning of the total R & D programme as well as financing the 'near market' part of it. The mechanisms of 'technology transfer' must be a subject for research in themselves.

Variability in consumer attributes is the key. Attempts to alter structures tend to founder on variability — to date they have failed to achieve an improvement on the extent of the existing chains' ability to continually deliver consistent eating qualities. In the shorter term 'quality assurance' schemes provide our best hope for improvement; in the longer term a combination of structural change and technological advance (both genetically and in the meat plant) will surely be necessary.

Expenditure on food has been declining as a percentage of total consumer expenditure, while expenditure on meat has been slowly decreasing as a percentage of total food expenditure — although that figure remains over 25%. The latter trend is a reflection of relative retail prices rather than volume changes.

In 1986, UK consumers spent over £8 billion on meat and meat products for domestic consumption (about £4·9 billion on fresh meats and £3·4 billion on processed meats); allowing for the 20% or so of production which gets utilised in the catering sector, the total outlay must be over £10 billion or some 4% of total consumer expenditure.

The proverbial Man from Mars might be forgiven for deducing from a study of our newspapers, magazines, radio and television items that meat eating is in rapid decline in Britain, In fact, per capita consumption has been steady in the 1980s and, against a background of adverse publicity has actually crept up in the last four years (Table 1).

Does this mean that most people are eating rather more, to compensate for reduced consumption among a small 'concerned' segment of the population? Most market research on this issue suffers from being based on the *claims* of sampled respondents as to their attitudes and habits, rather than purchase and consumption data, and the responses are sensitive to the form and manner of the questioning.

Two sets of survey results on vegetarianism demonstrate this. One, carried out each year for the Realeat Company (makers of Vegeburgers, etc.), has shown a small annual increase in vegetarian adults (from 2% in 1984 to 3% in 1987) and also in those claiming to avoid red meat (from 2% in 1984 to $3\frac{1}{2}$% in 1987) (Table 2).

But a second syndicated survey, reporting every quarter, has never shown more than 1% of vegetarians, with less than half the population claiming to eat less meat than they used to (Table 3).

Certainly many of us are eating less butchers' cuts of beef and lamb at home — but this reduction has been more than made up by growth in catering purchases, by greater use of meat in processed forms — and, of course, by the switch to lower cost poultry and pig meat. And the increasing range of alternative foods — including vegetarian meals — is bound to have an effect on the purchasing patterns of most families.

A marked trend in the total British market is in the growth in domestic production and exports and the decline in imports (Table 4), but this growth, of course, has been sustained, at least as far as beef and lamb is concerned, by substantial EEC support measures and border protection. In 1986/87, about 15% of beef producers' returns derived from the various support measures — a figure certainly in excess of the net margin achieved.

The key trend in other European countries, however, has been in the growth of consumption over the last twenty years (Table 5); in the EEC,

TABLE 1
Per Capita Consumption of Various Meats in UK (kg Carcase Weight)

Years	Beef	Lamb	Pork	Bacon/ham	Offal	Poultry	Total
1950–54	17·6	9·8	4·2	10·3	1·7	2·5	46·1
1955–59	22·7	10·6	8·1	10·9	2·1	3·8	58·2
1960–64	22·4	11·1	9·4	11·5	2·4	6·5	63·3
1965–69	20·9	10·5	11·0	11·4	2·4	8·8	65·0
1970–74	21·1	9·0	11·9	10·8	2·0	11·3	66·1
1975–79	22·1	7·4	11·3	8·9	2·2	12·4	64·3
1980–84	19·1	7·1	12·8	8·6	2·2	14·3	64·1
1985–88	18·9	6·6	13·2	8·1	1·8	17·2	65·8

Source: MAFF, MLC.

TABLE 2
Responses to Question Concerning Meat Eating Habits From Random Sample
of Adults in Four Years 1984–7 (percentages)

Claimed meat eating habit	1984	1985	1986	1987
'... avoid red meat'	1·9	2·6	3·1	3·6
'... vegetarian ...'	2·1	2·6	2·7	3·0
Each sample — approx 4 000				

Source: Gallup for Realeat Co.

TABLE 3
Responses to Question Regarding Meat Eating Habits from Random Sample of
Adults Taken in Eight Quarters from 1985–7 (percentages)

Claimed meat eating habit Year: Quarter:	'85 (4)	'86 (1)	'86 (2)	'86 (3)	'86 (4)	'87 (1)	'87 (2)	'87 (3)
'... as much as ever ...'	48	47	46	45	47	50	47	46
'... less than used to ...'	41	43	43	45	44	41	42	44
'... only rarely ...'	6	6	7	7	7	7	7	7
'... no meat but fish ...'	2	1	2	2	1	1	2	2
'... vegetarian ...'	1	1	1	1	1	1	1	1
'... vegan ...'	—	—	—	—	—	—	—	—
Each sample — approx 1 000								

Source: National Health Survey.

TABLE 4

Trends in Meat Production (P), Imports (I), Exports (E) and
Consumption (or Net Supplies (NS)) in the UK ('000 Tonnes)

Years		Beef, veal, lamb and pig meat	Poultry	Total
1970–5	P	2 269	635	2 904
	I	1 040	9	1 049
	E	95	2	97
	NS	3 211	643	3 854
1987	P	2 540	978	3 518
	I	752	81	833
	E	338	44	382
	NS	2 992	1 008	4 000

Source: MAFF, MLC.
Note: Net supplies also takes into account opening and closing stocks.

only Portugal now has a lower per capita meat consumption than the
UK. This, we believe, is primarily a reflection of the historically low
priority attached to food and to quality in food by British consumers,
who have been transfixed by price considerations and who have used
their increased affluence in other ways. However, there is clear evidence
that the relative importance attached to 'low prices' and 'quality' is now
changing in some segments of our population.

THE FOOD CHAIN AND STRUCTURE

The 'food chain' for meats is uniquely extended. It is a long way from the
decisions of the Limousin breeder on the choice of a bull, or of the
Welsh hill farmer concerned with which ewes to cull, to the decision of
the housewife whether to buy a pack of frozen burgers or a couple of loin
chops. If the loin chops turn out to be tough or the burgers disappear
under the grill — who is to blame? The traditional answer from the
industry is the cook!

But even under standardised cooking, the variations remain. We
have to accept that the extended chain creates many problems for
quality improvement and quality control, for successful marketing
against the competition of simpler foods and foreign supplies, and for

TABLE 5
Per Capita Consumption of Red Meat and Poultry in Countries of the EEC (kg Carcase Weight)

EEC member states	1960–69			1985		
	Beef, veal, lamb and pig meat	Poultry	Index	Beef, veal, lamb and pig meat	Poultry	Index
France	59·5	10·4	101	71·5	17·9	129
Belgium/Luxembourg	53·3	7·6	88	72·3	15·3	126
West Germany	61·7	6·1	98	84·1	9·7	135
Irish Republic	52·9	7·2	87	62·4	16·9	114
Denmark	50·4	3·7	78	71·8	11·0	119
Italy	29·3	8·4	55	57·1	18·0	108
Netherlands	47·0	3·9	74	61·0	13·7	108
Greece	27·1	4·5	46	56·6	15·7	104
UK	61·4	7·6	100	53·4	16·0	100

Source: MLC.
Index = Country as per cent of UK figure for total meat consumption.

'technology transfer'. And inevitably it creates problems for research planning.

A degree of frustration with the red meat sector can be detected in committee considerations of research priorities. These start from the position that food will need to be increasingly processed in innovative ways, that consumers are becoming more concerned with consistency and quality than price, that there is a need for greater variety and new products and that any growth is going to be export led.

Preference for research funds is likely to be given to food industries which fit in with these criteria, who have generated sufficient profits in recent years to be able to contemplate significant research investment at the applied end of the spectrum, who have shown themselves to be research-orientated and receptive to new technology, and who employ staff able to discuss research problems with scientists on an equal basis. The basic research problems appropriate to such industries and their next phase of development are likely to attract more government funds.

> The sponsors should take account of the willingness of various industry sectors to adopt the results of R & D when deciding how to allocate resources. (*Priorities Board, Second Report, 1987.*)

The challenge to the meat sector is clear and the penalties of not accepting it could be severe.

To the critic, the meat industry might appear to be poorly structured, traditional in approach, craft-based, not research-oriented, reluctant to accept new technology, still dependent on anachronistic marketing methods — such as the livestock auction market — and with its processing arm built around traditional formulations produced at the lowest possible cost for a mass domestic market not notably willing to pay for quality.

Although one can argue that other sectors also have their problems, at the heart of the problem in the meat sector is the historical level of profitability. And a major factor holding down profits is the structure of the industry. The price consciousness of consumers, the strength of the high-street multiples and their increasing demands, the small unit size of both producers and first-hand buyers and slaughterers, the continued reliance of some manufacturers on imported raw materials, the failure of producers to achieve dominance for their co-operatives (as have some of our more technically advanced competitors), and perhaps most important of all the genuine difficulties in offering the consumer something unique or distinctly different, all combine to make this a difficult industry within which *to finance* change.

The meat industry has not been short of advice and assistance from MLC and others about the need to recognise the influence of changing life-styles on consumption patterns and therefore demand, on the need to represent red meat cuts to today's consumers in lean, convenient, waste-free, ready-to-cook portions, to achieve guaranteed consistency in eating quality, to meet the challenges created by the growing importance of the multiple grocers and their wish to push cutting and packaging back to the abattoir, to innovate in processing, to control the 'cowboys' who undermine the industry's image, to support an export drive, to meet the needs of those consumers concerned about method of production as well as those prepared to pay for quality in a more traditional sense. And, indeed, to recognise the need for research, both basic and applied, and for industry to be prepared to contribute to its costs. All this is reiterated and re-emphasised in the MLC Corporate Plan which has been widely welcomed by industry organisations; but the challenge from government that industry should fund more of the 'near-market' research designed for its benefit (reflected in the Plan) is still meeting with some resistance.

There is movement in the meat sector — small it must be admitted — in all the respects I have outlined and it is bound to be accelerated by the packaging revolution bringing the meat pack supplied by the meat plant even nearer to the standardised grocery product — in principle if not in practice! But these developments are taking place tentatively within the existing competitive environment and structure of a vast commodity-based industry, few of whose participants have made sufficient profits over the last 10 years to fund re-investment adequately, and who face shrinking levels of market support with its implications for supply levels and, therefore, volume throughput in meat plants. These are the businesses who must now dig into their pockets to fund *research for change*.

CRITICAL RESEARCH

It is easy to criticise the meat industry for being slow to take up the results of research. But just how far has research progressed with the key problem of understanding and controlling variability in meat? Has research got to the point when a clear technology can be laid down to achieve consistency, capable of being adopted by progressive meat traders in today's competitive environment?

For example, consider the problems of minimising the variation in the eating quality of beef, even when it is prepared at the meat plant for retail sale in a thoroughly modern format (lean minces, cubes of lean meat for slow cooking, kebabs, stir-frys, portion-controlled steaks, possibly part-cooked), in advanced packing, without resorting to the conventional techniques of 'processing' such as adding water and additives, tumbling, massaging, flaking and reforming. Can we achieve high repeatability in advanced preparation cuts, as opposed to processed products, when the beef carcases broken down for these purposes will probably have been bought opportunistically, probably by eye judgment of the live animal with little information on genetics or management or, indeed ownership, and conventionally slaughtered, dressed and chilled? Is reduced variability possible without a wholesale upheaval in industry structure and trading practice?

Why does the multiple retailer, in specifying his requirements to the abattoir/packer, generally not stipulate particular breeding, management and feeding and require evidence on ownership, mode of transport to the abattoir and handling practices there, before and after slaughter? He may seek to do so, at least in the form of guidelines, but insistence on such a tight specification would impose problems of control and probably dry up his supply immediately. The premium required by everyone involved to change their existing trading practices and habits would add up to a total on-cost which would render the final product uncompetitive in the market. (Indeed the multiple is likely to be looking for a *lower* price than the norm, reflecting, they would argue, the volume and regularity of their requirements, rather than a premium.)

The industry needs to find ways round this impasse. Small incremental changes in practice before and after the farm gate — individually not detectably beneficial — can, we would argue, accumulate to show a real improvement. Under current structures, quality assurance schemes can, we believe, provide the necessary discipline and incentives and should be pursued with vigour.

There is, as yet, no clear evidence that supply controlled by tight specification of breed type, feeding system, selection for slaughter, etc. *does* provide a more consistent product than can be achieved by careful selection from a large pool, coupled with tight carcase specification and abattoir handling procedures. Would any improvement in consistency or quality be sufficiently clear and demonstrable to consumers to earn a premium from them? The need to demonstrate this must be high on anyone's list of research priorities, since without it the need for quality assurance schemes in the shorter term, and structural changes in the longer term, will be difficult to sell. It is by definition a 'whole chain' research problem.

THE SCALE PROBLEM

Small scale enterprises buying and selling locally may get close to this ideal of specification, but major national multiples with perhaps a 5% market share would have a formidable task in identifying specific production units as the source of their requirements. For example, 5% of the retail beef market is about 120 000 clean cattle equivalents — or more likely 180 000 cattle, because the typical multiple does not expect to sell the whole carcase in balance. Bearing in mind the number of finished cattle marketed by the average beef finisher each year, this means establishing contact with some 5000 finishers, who will obtain their raw material from an even greater number of calf or store cattle producers — again with the purchasing decisions being almost entirely based on eye judgements. But this assumes that all flows evenly during the year; of course, it does not. Particular producers will market at particular times appropriate to their production systems — in fact larger numbers of finishers than 5000 will be involved in order to achieve a full beef cabinet for the multiple retailer throughout the year. Difficult though it is, some multiples are improving direct producer relationships, but *not* for direct trading.

It is the meat wholesalers'/abattoir operators' job to match the fragmented and variable supply with the demand and there is no suitable volume alternative. Producer groups have sought to make a contribution, stimulated by an understandable wish to get nearer their market and to earn a premium by meeting a particular retail buyer's requirements more closely, but they tend to lose control once they seek to expand. From the abattoir operator's point of view, he needs to be convinced that a small number of contracted producers can actually provide him with greater consistency and continuity than his buyers can achieve by selecting from the large pool available daily in the markets and, further, that they will not be seduced from a contractual commitment by short-term alternatives.

When we control as well as practically possible the pre-slaughter handling, the refrigeration, the ageing time, the preparation and the cooking, we still get variations in eating satisfaction, whereby the stew can prove tasteless, the steak tough or the roast dry. Is the next step to attempt to tighten the animal specification or to look to our control procedures at the abattoir? Can we ever really control refrigeration to the extent necessary as long as we hang up 300 irregularly shaped and variable items together in a vast turbulent cold cavern?

In short, the main research problem is as it always has been, what is the extent of variability in eating satisfaction, what causes the variability and how can it be controlled? Can we get very far with conventional specifications which are, in any case, exceedingly difficult to control because of structure and trading practice, or must we await major technological change, either in the meat plant or in the breeding (genetic engineering to improve consistency)?

THE PROCESSORS

The manufacturer who uses meat as a raw material for meat products by comminution or other processes that change the structure, and who reforms the material into a new shape, is basically interested in cheap lean meat. He buys on the basis of visual lean (according to cut of origin 70–95%) and bacteriological qualities; he may be concerned to a degree with water binding capacity and connective tissue content — but if carcases from mixed sources are being deboned, trimmed and packed for a manufacturer's order, it is unlikely that whole boxes will suffer from a systematic defect in either of these respects and any localised variations will be lost in the grinding or flaking and mixing. Efficiency of trimming, closeness to specification, absence of foreign bodies and bacteriological status will dominate the buyer's concern in addition to price.

His interest in research into his raw material will be confined to 'how to make it cheaper, cleaner and leaner'. Beyond that he is concerned with problems of food processing technology in order to reduce his production costs and to achieve the product variety and qualities demanded by the consumer and articulated by the manufacturer's customers — the retailer, particularly the multiple retailer, and the caterer. The Gorsuch report has recently examined priorities in research in food processing and much of that analysis is relevant to the meat sector.

Livestock producers do not generally produce a raw material for meat manufacturers (except for bacon curers). They produce meat animals for the fresh meat market, knowing that certain cuts from certain carcases, or in some extreme cases, whole carcases, may be worth more as raw material for a manufacturer than they are as fresh meat cuts. He naturally hopes to maximise the proportion of production that has the higher value use. Even in the feeding of cull cows to increase saleable weight, for example, there is the hope that the carcase can be up-graded to the point when certain higher valued cuts can be 'robbed' and used in the fresh meat business, possibly following some mechanical or enzymic tenderisation process.

While it is understandable that planners should envisage a meat industry in which a higher proportion of meat is processed in ways which allow greater uniformity and variety, meat will remain a relatively expensive raw material for processors and hence in such products will always be a candidate for substitution by bio-engineered materials of plant origin. This same concern does not apply to 'advanced preparation' products, as opposed to processed products, where the meat is still recognisably meat and where basic concerns about variability in eating qualities still apply.

PRODUCTION COST

The consumers' appreciation of the qualities and variability of lean meat and of the range of meat products in form, taste and appearance is one of the two basic dimensions of the meat research problem. The other concerns basic costs of raw material production; the amalgam of performance and yield, usually summarised into the single trait of lean tissue food conversion. We want this reduced in order to offer consumers lean meat as cheaply as possible. Research has concentrated on this issue — with considerable impact on industry and benefits to consumers. And further progress will be made using new biotechnology and techniques based on new physiological understanding, as well as the more conventional systems approaches.

In social terms, however, there are two additional challenges. The first is that some consumers are concerned not only about price, suitability for purpose and taste — but also about *how* we produce for them cheap, lean, convenient meats and products. Their concern may be with animal exploitation or cruelty, unnatural conditions, genetic manipulation, growth stimulants or techniques and additives in food processing. The niche marketing opportunities that these concerns create will perhaps provide a lifeline for smaller traditional meat businesses unable and unwilling to join in the technological rat-race for the 'mass market'.

The second additional dimension is provided by the fact that enterprise choice, relative costs, profitability and even quality criteria are influenced by the political framework within which production and marketing operates and in particular by the CAP. The spectre of over-supply, which is due to misdirected support mechanisms, is influencing decisions throughout the industry, including decisions on R & D priorities.

These factors interact in interesting ways. For example, there always has been the possibility of an interaction between the drive for cheaper lean meat and 'quality', in the sense of eating quality. To this day, many meat traders will argue that fast grown, lean, youthful beef is of inherently inferior 'quality' to slower grown, older, fatter beef; however, today's consumers may vote with their money. And the pursuit of fast lean growth has undoubtedly had some effects on the 'qualities' of pig meat.

The production 'constraints' issue clearly interacts both with the pursuit of cheap lean meat and with 'quality'. In the developing social atmosphere, which is antagonistic to the deployment of 'science' to engineer cheaper consistent food products, how much do we need to redirect resources from performance optimisation within conventional systems, with few holds barred, to the achievement of performance improvements within constrained systems? The strength of social forces has to be judged and our domestic 'problems of bored affluence' have to be set in the context of world food production needs. These problems are more important and more emotive for animal and meat science than any other branch of food production.

Further, when evaluating the impact of 'constraints' (as with 'quality'), one has to ask not only does the consumer want something but *how much*? Is the premium obtainable from the market commensurate with the costs added by operating constrained systems? Do we lose one benefit we have painfully achieved when we go for another newer supposed benefit? For example, a multiple meat trader when offered 'free-range pork' sampled and liked its taste and flavour — but sent off the producer with the message 'Fine — we are interested, but by the way we want no more than half the fat of your samples!'.

Market management mechanisms influence the balance of production between species, between systems with the same species (extensive vs intensive, less favoured areas), costs of lean meat production and quality differentials. Regulations can have a rather direct influence on 'quality' — perhaps the best known example is the requirement that meat for intra-Community trade should have a deep temperature below 7°C before despatch; the rapid chilling involved was in danger of toughening the meat.

Market managers, and their political masters, have allowed the prospect of over-production to cloud their judgment of techniques designed to improve productivity and quality — notably growth-promoting hormones. They opted for an unenforceable ban rather than for a scenario in which meat produced without benefit of such growth promoters could compete for market share and for the necessary premium against meat produced with the aid of these stimulants among informed consumers. This is now facing research planners with a real dilemma; is it worth pursuing methods of efficiency improvement that involve feed additives, injection or other 'unnatural' practices?

CONCLUSION

The need is to help, by careful injection of government R & D funds, sectors whose advancement will benefit the national economy and/or whose degeneration would have severe social, environmental or balance of trade consequences.

At the same time, the sector must be assisted to overcome imperfections of structure, adaptability, attitude or awareness that hinder the conversion of research into improved practice or products, and encouraged to co-operate in the planning of the total programme as well as the financing of the 'near-market' part of it.

These thoughts on a complex subject will, I hope, provide a basis for discussion, particularly about how we work within a particular industry (with many imperfections) to identify the real problems at the applied end of the R & D spectrum, how we convince the sector that it is in its medium term interest to finance research for change and how we achieve adoption of results.

The latter discussion needs to come higher up the agenda — indeed the mechanisms of 'technology transfer' must be a subject for research in themselves. Not every meat trader is pig-headed or reactionary; most are quite prepared to contemplate change if benefits exceed costs, but if benefits are intangible (or not readily measurable), or if they accrue to others in the chain, or if costs include not only financial costs but loss of competitive position, then their reluctance to change is understandable and should be recognised in planning.

Variability in consumer attributes is the key. Attempts to alter structures tend to founder on variability — to date they have failed to achieve an improvement on the extent of the existing chains' ability continually to deliver consistent eating qualities. In the shorter term, quality assurance schemes provide our best hope for improvement; in the longer term a combination of structural change and technological advance (both genetically and in the meat plant) will surely be necessary.

Oils and Fats

A. CROSSLEY

Unilever Research Laboratory, Bedford, UK

ABSTRACT

80% of the world's production of oils and fats are used in food. Some years ago, the field appeared relatively static in terms of procurement, production and use. Today, major changes are occurring which challenge a number of disciplines. There is no lack of availability of fats. Change results from new consumer demands and from the availability of new technology — not least in the biological area. Low cost and precise manufacture is a major requirement.

Nutritional guidelines indicate a need for lower fat consumption and lower levels of saturated acids. New low-fat, and even zero-fat analogues of current products are appearing. Agricultural research provides the means of increasing productivity, and a wider range of raw materials via selection, cloning and DNA manipulation. New processes such as enzymic modification of fats are emerging.

The trends and the underlying technology are considered and suggestions made for technical areas for future development.

INTRODUCTION

World production of fats is currently *c*. 73 M tonnes of which *c*. 80% is used in human foods for edible purposes, the bulk of the remainder being used for production of soap, oleochemicals and animal feed. Figures for production for selected oils and fats are shown in Table 1.

The pattern of consumption of fats varies very much between countries. Moreover, the balance of fat consumption between 'oils and

213

TABLE 1
Structure of World Oils/Fats Market
(1985/86 Season)

	Million tonnes	
	World production	World exports[a]
Consumer products		
Butter	6·4	0·6
Lard	5·5	0·4
Olive oil	1·6	0·3
Total	13·5	1·3
Industrial type raw materials		
Linseed, castor, tung	1·3	0·6
Tallow	6·3	1·9
Total	7·6	2·5
Edible type raw materials		
Soya	15·7	6·1
Palm	7·7	5·3
Sunflower	7·2	2·1
Rape	6·3	1·7
Coconut	3·4	1·6
Palm kernel	1·0	0·7
Groundnut	3·2	0·6
Cotton	3·4	0·4
Maize	0·9	0·2
Sesame	0·7	0·1
Fish	1·5	0·7
Other	1·3	0·2
Total	52·3	19·7
Total of all oils/fats	73·4	23·5

[a]Seed and oil exports combined in oil terms.

fats' (largely vegetable in origin) and fats derived from the diet, often called 'hidden fats' (largely animal fats in the West), also varies by country. The balance is important in considering nutritional implications (see below).

Forecasts indicate a gradual expansion of production to the end of the century — perhaps to *c*. 105 M tonnes. Within this there will be a marked increase in the proportion of palm oil, which already (in 1985/6) provided 27% of world exports of edible oils. The proportion of marine oil is relatively low at *c*. 2% of the total. No shortage of oils and fats generally is foreseen in the next decade — although the possibility of crop failure always exists, giving rise to short term perturbation of the picture.

SUPPLY OF OILS AND FATS — TRENDS

Major changes in types of oilseeds produced are not anticipated, but, within this scenario, important advances are likely. Palm plantlets can now be provided using cell culture techniques, enabling selection for oil yield, disease resistance, etc. There will undoubtedly be a steady increase in oil yield via selection of prolific strains for cloning. It has been claimed that yields could double eventually. Similar techniques can be applied to other species. The technique has also been used to provide cloned palms with novel fatty acid compositions, e.g. low or high in linoleic acid. This range of possibilities is as yet little studied.

Conventional plant breeding is also providing modified oil compositions. The best example is the virtual elimination of erucic acid from the glycerides of rapeseed oil. Safflower oil, and more recently sunflower oil — traditionally rich in linoleic acid — can now be obtained commercially, with linoleic acid largely replaced by oleic acid. Variants of both sunflower and rape are being produced which are suitable for growing in a wider range of environmental conditions, e.g. sunflower in the UK.

Among the less conventional oilseed crops, the tree, *Sapium sebiferum*, which yields a hard 'tallow' rich in palmitic and oleic acids, is fairly widely grown in China and is proposed for growing in areas unsuitable for other oilseed crops. Lupins are being developed as a source of liquid oil. Cuphea is proposed as a source of oil rich in shorter chain fatty acids (C_8 to C_{12}).

Overall, one cannot avoid the view that what has been a rather static scene is changing rapidly. Optimised and new crops are being developed and new propagation methods will accelerate the trend. In the long term, genetic engineering of plant DNA, facilitated by cell culture techniques, should allow even greater facility to manipulate oil composition.

Microbial production of oils and fats has been studied. Feasibility is well established and yields from microorganisms dramatically increased. However, the costs of the process are unlikely to allow it to be economic, except perhaps in countries where the cost of biomass is very low, or for the production of speciality lipids.

It is not possible to consider the future of oils and fats without considering the principal by-product — protein for animal feed. It can be equally important to optimise this component as to optimise the nature of the oil — though less researched. Toxic or anti-nutritional factors in the 'meal' are particularly important. The use of rape meal has been severely limited by the presence of such factors. New varieties of 'double zero' rape are beginning to overcome the problem. Proposed EEC legislation ($<20\,\mu$moles glucosinolate component/g) would eliminate many of the rape varieties currently being grown. Further development of 'double zero' varieties is urgent. Similar considerations apply to lupin and to a lesser extent soya. The use to which the protein is put is also important. Much seed protein has traditionally been used for dairy and beef production. With declining dairy herds, the application to pig and poultry feeds of some of the newer meals becomes more important. This is less well understood and has been limited in rape by the glucosinolate content.

Finally, mention needs to be made of 'Fat-Replacers'. 'Sucrose polyesters' have been developed recently which are not metabolisable but have many of the characteristics of conventional oils and fats. Presumably other chemical species could provide similar effects. Jojoba oil (rich in long chain alcohol monoesters) is grown commercially in limited quantities and has been proposed for this purpose. It seems unlikely that such materials will match conventional fats in cost, but could nevertheless find wide application.

Conclusions — Supply of Oils and Fats

Technology is rapidly developing to enable modification of oilseed yields and composition, and to provide wider growing regimes. Fat substitutes may be a new area ripe for exploitation. It is very important that both growers and users clarify their targets in order to provide directions for research, e.g. optimum physical and chemical properties of the oil. The needs for protein feed should be considered simultaneously. This is a fruitful area of research which should be given considerable emphasis. Many of the targets are 'global' rather than specific to a particular product target or one country. Funding for such research should therefore be given strong support from governments.

PRODUCT APPLICATIONS

The requirement for R and D will be very dependent on product needs. It is important therefore to discuss this aspect early in this paper.

Major applications for oils and fats are frying and salad oils, margarine, cooking and bakery fats, and confectionery fats for chocolate and other goods. There is a wide variety of other edible uses of oils and fats — but in many cases these can utilise products and technology developed for the major uses.

Frying Oils

In Europe, there has been a major increase in the use of oils, rather than fats for frying. The consumer demands that oil should not be 'cloudy', even at low temperatures. The major technical requirement is for optimum stability both at ambient and high temperatures. To minimise cost, ideally all liquid oils should be freely usable.

The research requirement is to study the oxidation process and to relate it to realistic product assessment. The value of antioxidants, both synthetic and natural, needs close definition. Proposed antioxidants must be fully acceptable in use toxicologically. The elimination of more highly unsaturated components provides a means of enhancing stability. Oils with maximum linoleic incorporation (for nutrition) and also oils with maximum oleic incorporation (for stability) would be appreciated by the industry. Flexibility and interchangeability of use of oil types is valuable. Rape and soyabean oils should be provided to quality standards equal to sunflower and maize by minimising the level of linolenic acid in these oils, thus improving oxidative stability. Liquid oil from palm oil sources should be sought, e.g. by seeking varieties containing more unsaturated acids and also by improved physical separation of olein and stearin (see below). Palm oil with low levels of colour and improved colour stability on processing is desirable.

Margarine/Spreads

Margarine is an emulsion with fat as the continuous phase. The continuity of the fat provides the relatively long microbiological stability, typical of these products. The supermarket shelves reflect the change occurring in this market. Products with different fat levels (as low as 30–40% fat) are appearing, products with high levels of polyunsaturated acids are taking a major share and melanges of butter and vegetable fats are becoming available.

These changing product types demand background support from R and D for their future development — and of course change the nature of the manufacturing operation.

The requirements for individual oil/fat components are complex. Components are invariably blended. The blend should provide: (a) the physical characteristics required in the final product, including appearance, spreadability (often at refrigerator temperatures) and oral response; (b) the ability to maintain these characteristics on storage; (c) in many cases the ability for use both for spreading and cooking; and (d) the ability to be processed satisfactorily into the final emulsified product. The product use may impose further demands, e.g. maximum linoleic component, minimum saturated acids, etc.

There are further requirements for suitably functional minor components — flavours, emulsifiers, agents for pH adjustment, and thickening agents for the aqueous phase. Whilst important, these are outside the scope of this paper.

It is clear, considering the types of oils and fats available, and the product requirements, that the need for relatively hard fats cannot be met from natural sources directly. The shortfall is provided by chemically or physically modifying the oils to provide harder fats — largely using hydrogenation (see below). This then provides a large variety of potential raw materials available for blending.

A wide range of glyceride types is present in the final blends. It is the crystallisation characteristics of the blend which largely determine its performance. The solid fat in the product, in the form of crystals, plays a key role in determining the hardness at various temperatures and also the integrity of the emulsion structure. The 'rules' which govern this behaviour and the relationship to fat composition are expectedly complex. Glycerides can interact physically (eutectic formation) and various polymorphic changes take place in the processing sequence (e.g. during the chilling stages of manufacture). Moreover, the formation of the final product structure takes place over a period of days rather than minutes. Manufacturers use their accumulated data to formulate products on the basis of linear programmes, selecting the cheapest formulations to achieve a standard performance. There is no doubt that research is still needed in many aspects. Detailed studies of crystallisation kinetics and identification of the resulting product structures using the most modern techniques will provide new insights.

Physical measurement, and wherever possible in-line sensors, can be valuable in controlling the complex manufacturing operation. New product needs, e.g. margarines with high levels of linoleic or minimal saturated acid components, impose further technical constraints that need a detailed research response. The trend to low-fat products imposes a new set of technical problems.

Bakery Fats

The technical needs for bakery fats vary with the specific application, e.g. puff pastry, pie pastry, cakes for both kitchen and industrial application. A small amount of high-melting fat component is needed and the properties of the remainder can be critical. 'Creaming' properties rely on the ability to incorporate air in an emulsion. Emulsifier type and fat crystal morphology are critical. Much of the technology developed in the margarine area is applicable.

Confectionery Fats

Chocolate and confectionery fats are typified by 'sharp melting': they contain a large percentage of solid fat at lower temperatures, but melt fully in the mouth. Among natural fats, cocoa butter is almost unique in providing this function. Common fats, particularly palm oil, can be fractionated to provide a sharp-melting characteristic (see below).

The melting performance of cocoa butter and related fats in terms of glyceride composition is associated with their 'purity' relative to margarine fats, which contain a wide range of glyceride types. One result of this 'purity' is complex crystallisation behaviour. The stable polymorphic form, i.e. fat crystal packing, is β, unlike margarine fats, which normally stabilise in the β' form. The practical consequence of this is the complex and time-consuming cooling sequence, which is used in the manufacture of chocolate in order to provide the stable form. If this sequence is not followed rigidly, the chocolate is soft and suffers from the well known 'chocolate bloom', due to subsequent polymorphic phase change in the final product. Full understanding of these changes is not available, yet is particularly important in designing alternatives to cocoa butter.

Another category of confectionery fats exists based on palm kernel and coconut oil (again using a fractionation process). With these materials, crystallisation problems are less serious, but unfortunately these products cannot be used in normal chocolate (containing cocoa butter), due to eutectic effects in the mixing of glyceride types.

Conclusions — Products

Changes in product types, e.g. in frying oils and margarine-like spreads, largely resulting from new health trends, pose new problems in product formulation, product stability and processing. Lower-fat spreads have an increased requirement for a well characterised and reliable chill distribution.

Cocoa butter is the traditional confectionery fat. Its use is well known, although further detailed understanding of its behaviour during processing is still valuable. There is also a need to develop a wide range of relatively cheap, sharp melting 'equivalent' or 'substitute' fats for a variety of purposes, including use in normal chocolate. The production of these by physical or chemical means is under study (see below). A range of products is commercially available.

Detailed studies of the kinetics of fat crystallisation are essential to support all these areas. Such studies should be linked to studies of final product structure. Process sensors can be improved or developed to optimise the processing, e.g. to monitor the levels of solid fat in process, and the development of a tempered chocolate mass prior to setting.

There is a need to provide quantitative links between consumer perception and the measurable parameters of the products and their processing. This form of consumer research, usually termed Consumer Research Guidance, is a general need in food product development and requires further development.

NUTRITION

The effects of both level and type of fats in the diet have attracted world-wide attention. Governments have made a number of recommendations. In the UK, several Committees have made statements which have received wide publicity. Many aspects are still under debate. Least contentious is the proposal to reduce overall fat consumption from 40% towards 30% kcal of the diet (together with overall energy reduction). There are strong views that lowering saturated acids and increasing polyunsaturated acids is important in relation to coronary heart disease. There is a growing feeling that the presence of monounsaturated acids at the expense of saturated acids can provide advantages in particular types of diet. Recently, views have been expressed that the so-called ω-3 fatty acids have particular advantages — for example, EPA (eicosapentaenoic acid) and DHA (docosahexaenoic acid) in fish oil.

The problem of adding unhydrogenated fish oil to products would require further study in terms of organoleptic acceptability.

There has been much debate on the nutritional effect of hydrogenated fats, which are substantially modified in chemical terms. The current view is that they present no problem.

Some years ago, a number of studies implicated erucic acid (C_{22}) in rape oil in heart disease. Whilst final proof was difficult to achieve, the effect was virtually to eliminate 'high erucic rape' from the diet, to be replaced by the current products virtually free from erucic acid.

Conclusions — Nutrition

There is a clear need for continuing biological assessment of these effects. However, the data already available and the statements by governments make it important that products are available in order that consumers can implement proposals. The 'hidden' fat in a typical diet is less easy to modify and the fat composition of foods such as margarine and frying/cooking oils is therefore particularly crucial. Lowering the average fat consumption is clearly advisable. There are strong indications that saturated acids in particular should be reduced — to be replaced by more unsaturated oils. Oils containing linoleic acid ('PUFA') should represent a higher proportion of the fat in our diet, whilst reducing total fat intake. Changes to food types to implement nutritional directions require substantial development.

OIL REFINING

Oils and fats are neutralised with alkali to remove fatty acid, bleached to remove colour and other materials, and then deodorised under vacuum at high temperature. 'Physical refining' (treatment of oils at high temperatures with steam to remove free fatty acids, odours, etc.) is becoming more important with considerable financial benefit, as it eliminates a separate neutralisation stage. Much development has been carried out (and will continue), but with very specific objectives — usually cost reduction. There are a number of areas where modern sophisticated sensors can be explored, in particular to check purity and quality of oils at all stages of processing.

Conclusions — Oil Refining

Technological work to cheapen processing is worthwhile. Sensor development can improve efficiency.

OILS AND FATS MODIFICATION

Hydrogenation

Chemical hydrogenation of oils and fats, usually using a nickel catalyst, is a well established process. Oils and fats can be hydrogenated to various degrees to yield a variety of products. Melting point is raised as double bonds are reduced, and also (equally importantly) high melting *trans* isomers are produced. Without the hydrogenated fats many of today's products could not be made. World production of these fats well exceeds one million tonnes per annum.

A major objective of the process is to provide the harder fats needed in products. Also important is to minimise more highly unsaturated acid components. Thus unhardened fish oil (not normally used in products), when hydrogenated, loses its typical fishy off-flavour; soya and rape have improved stability when the linolenic acid (triene) is removed. Some targets still remain e.g. improved 'selectivity' of hydrogenation in relation to the more unsaturated species. Process control, e.g. using sensors, also deserves more attention. In addition, biological hydrogenation, which in theory appears to be feasible, requires consideration, but the research objectives need clear definition, e.g. to replace chemical hydrogenation for consumer appeal *or* to provide a different product, i.e. chemical compositions not obtainable by the conventional process. However, in view of the difficulties in developing a biological process at a reasonable cost, one must question the value of this approach.

Fractionation

Fats have been fractionated using crystallisation for many years and for a variety of purposes: to provide a liquid oil from palm oil, fractionation of palm kernel oil to provide a 'sharp melting' confectionery fat, and fractions of animal fats (including butter) for a variety of purposes. A particularly interesting application is the fractionation of palm oil from solvent (usually acetone) to provide a liquid oil (60%), a 'sharp melting' confectionery fat (30%) and a hard fat with bakery uses (10%). The 'sharp melting' component of palm oil is particularly interesting, as it is physically compatible with the glycerides of cocoa butter and therefore is a valuable component in good quality chocolate.

Conventional fat crystallisation is carried out in the presence or absence of solvent (in one particular case in the presence of detergent solution), followed by separation by filtration, hydraulic pressing or centrifugation (the latter particularly for the detergent-based process).

Other processes have been proposed, e.g. liquid extraction, supercritical extraction, and adsorption systems.

The oils and fats users are making increased use of fractionation processes (but there is still a need to brief researchers precisely on needs). This is a fruitful field for research and development.

Interesterification

Chemical interesterification

Fats and blends of fats can be modified by chemical interesterification, using trace amounts of alkali (in various forms) as catalyst. The fatty acid residues are equilibrated between the various glycerides, thus changing the physical nature of the product. It is also possible to 'direct' the process by simultaneously crystallising to provide a 'non-equilibrium' product mix ('directed interesterification'). Melting profiles and crystallisation behaviour are modified. The products have a variety of uses — either as components of cheaper formulations or products with improved performance. The processes are well known and require little further development. An example is given below:

```
 ┌─ oleic            ┌─ stearic
 │                   │                 Na⁺          6 glyceride
 ├─ oleic      +     ├─ stearic         ⇌            species
 │                   │                 100 °C
 └─ oleic            └─ stearic
   Triolein            Tristearin
```

Biological 'specific' interesterification

A new biological process has recently been developed. This uses a fungal lipase as catalyst and operates under mild conditions with few by-products. This reaction occurs in the virtual absence of water, although traces of water are needed to activate the enzyme. If a 'random lipase' is used, the effect is identical to the chemical process. If, however, a 'selective' lipase is used, other effects can be achieved. Many lipases are 'position-specific', promoting interesterification only at the 1- and 3-positions of the glyceride molecule. Moreover, it has also been found that glycerides and fatty acids or simple esters can be used as co-reactants. For example, the glyceride triolein can be reacted with stearic acid to provide the glyceride: 1-stearic—2-oleic—3-stearic (St—O—St) (cf. the chemical method shown above).

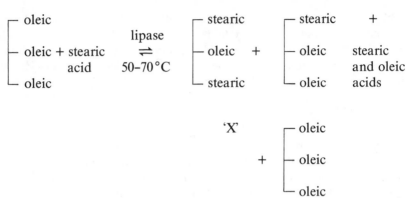

This type of product (the glyceride marked 'X' above which is readily separated from the mixture by fractionation) is very similar chemically to cocoa butter and can be used as a substitute. It cannot be provided easily by chemical processes. In application as a cocoa butter equivalent, it will normally be mixed with a sharp-melting palm fraction (see above). Intensive development is taking place. There are many other applications of this new (bio)technology in the fats area, e.g. replacement of conventional interesterification using a 'random' enzyme, inclusion of unsaturated acids in saturated fats, etc. Moreover, lipases exist with other specificities. These are also being studied and novel lipases sought. The use of lipase-mediated biotransformation extends beyond triglycerides, e.g. to the production of phospholipids and of mono- and diglycerides.

Conclusions — Modification

Hydrogenation is already intensively researched. Some objectives still remain, such as improved 'selectivity'. Biological hydrogenation needs careful evaluation.

Fractionation is technologically fairly well developed and is in large-scale use for fractionated palm oil, tallow and other oils and fats — both in a solvent and non-solvent mode. It merits detailed study both of the crystallisation process and its effect on the subsequent separation stage. Novel processes merit further study.

Chemical interesterification needs no further research, but *biological interesterification* merits intensive study for both modification of fats and other lipid systems.

ANALYSIS

Fats are a complex mixture of triglycerides. They may also contain appreciable amounts of minor ingredients — gums, tocopherols, diglycerides, monoglycerides, phospholipids, etc. Product properties are frequently very sensitive to composition. Chemical methods of analysis need to be precise — and, for process control, rapid. Analytical methods are not always adequate and frequently require specialist skills. New developments in capillary gas–liquid chromatography, in high pressure liquid chromatography, etc. are directly applicable. There should be a move from laboratory quality control in the application of these techniques to process control-on-line where appropriate. The need is therefore for rapid and robust methods at reasonable cost.

Analytical methods for control of possible contaminants, e.g. pesticides, are very important. In addition to chemical analysis a number of physical methods of analysis are under intensive development — often in areas other than fats. These include nuclear magnetic resonance, differential scanning calorimetry, and microscopic techniques, including laser and X-ray confocal techniques. All of these need further development for fats-based product systems.

Overall, there should be a target of more precise, simple and rapid means of chemical analysis and physical characterisation, enabling better evaluation of raw materials and better optimisation of processes, leading to a more predictive means of linking composition with product performance.

Conclusions — Analysis

There is a rapid development of both chemical and physical techniques. The applicability to oils and fats should be studied. More precise and rapid techniques are highly desirable.

SUMMARY

R and D in the area of fats is intense, as shown by the volume of publications and patents appearing. A major impetus has been the greater understanding (though far from complete) of the dietary role of fats — with new products emerging to satisfy these needs. Changes are also taking place in the supply of oils and fats — evident in Europe by the extensive plantings of rape and sunflower. New and more cost-effective plant varieties are being provided by conventional plant breeding and also by cloning in cell culture. In the future, genetic manipulation of plant DNA is likely to provide more diversification and more cost-effective varieties. This area requires the closest collaboration of all interests — including government — to provide clear targets and achieve optimal development. A new raw material, sucrose polyester, produced chemically, is being developed as a zero-calorie food component in place of the conventional oils and fats.

Among many developments in oils and fats processing, a biochemical route stands out. Fats can be modified by interesterification with other fats and fatty acids, using a mild enzymic process. The route is 'selective' and should lead in some cases to more cost-effective production, in other cases to novel fats for product application.

In the fat-based food products the major trends are towards products with lower fat levels and/or containing substantial quantities of more unsaturated oils — both primarily for health reasons. Products containing higher amounts of unsaturated oils require attention to protection from oxidation (better antioxidants/new oilseed varieties). Low-fat analogues of fat-based products require greater care in microbiological preservation and use of a reliable chill distribution system.

Low-fat 'halvarines', which are emerging in many forms in the supermarket, pose a major challenge to scientists and engineers to provide products (and processing regimes) yielding product structures acceptable to the user (and improved techniques to link product structure with consumer perception). The provision of a stable fat-continuous emulsion of this type with relatively low amounts of fat also requires particular attention to the nature of the aqueous phase components.

Modern production control methods are applicable to oils and fats processing. The use of computers is essential, e.g. in factories and production modelling and in computer-integrated manufacturing (CIM). The variable nature of the raw material can be dealt with by linear programming techniques. However, to optimise these approaches, there is a need for sensors to monitor the key parameters — nickel in oil during hydrogenation, water in emulsions for margarine manufacture, development of fat solids during processing, etc. The development of oils and fats manufacturing with low labour costs is becoming a reality.

Overall, oils and fats technology is in a state of rapid development with a wide range of inputs needed from a variety of sources, but with the necessity of coordination of objectives and with governments playing an important role.

Cereals

J. D. SCHOFIELD

Flour Milling and Baking Research Association, Chorleywood, UK

ABSTRACT

Cereal-based foods form an important element of the UK diet. Dramatic changes have occurred in raw material supply, processing technology and markets, and R & D has either helped processors to cope with change or helped to bring it about. Change continues, creating further requirements for R & D. Cereals Sector companies invest substantially in R & D, and the industry looks to the public sector for support of both basic and applied research to underpin its own effort, which is usually directed towards clearly defined commercial objectives. R & D requirements are similar, in broad terms, to those of other sectors and include: (a) more precise understanding of raw material composition, changes during processing and the physics and engineering principles of unit processes; (b) development of appropriate sensor and/or artificial intelligence systems for application in process control and automation; (c) development of improved measures to ensure maintenance of product quality during storage and distribution, and (d) provision of information to enable consumers to select from a wide variety of cereal-based foods as part of a nutritious, safe and properly balanced diet.

INTRODUCTION

Cereals have traditionally been the mainstay of the nutrition of major civilisations. Even today, with the plethora of food products available, they are still major sources of nutrients in the average diet. In the UK, cereal-based products provide about 30% of the energy, 25% of the

227

protein, 50% of the carbohydrate, 40% of iron, but only 10% of the fat in the diet.

As a result of their importance, cereal-based products, particularly bread, have always had a high political and social profile. Furthermore, factors such as changes in the cereals supply position, changes in consumption patterns and changing economic, political and social pressures have all had significant impacts on the requirements for research and development and technological innovation in this sector. The UK cereals industries have adapted well to these changed circumstances, and this is due in no small part to the contribution from R & D.

A total of 5·3 M tonnes of all grains was used directly in foods for human consumption in the UK in 1984/85. Of this total, 4·6 M tonnes were wheat for milling and 0·28 M tonnes of wheat/maize for breakfast cereals. Most of the oats (0·145 M tonnes) was also used for breakfast foods. A further 0·87 M tonnes of maize and 0·2 M tonnes of wheat were used for starch/glucose/gluten/etc. production.

Household expenditure on cereal-based products (excluding alcoholic beverages) was over £4 milliard in 1985, representing about 14% of total expenditure on food. But substantial amounts of cereal-based products are also consumed outside the home in various types of catering establishment, and the total retail sales value in this sector in 1985 amounted to about £4·5 milliard, i.e. similar to that for dairy products.

Trade in finished cereal products between the UK and other countries is significant, although the value of exports (c. £120 million) and imports (c. £100 million) is relatively small compared with total domestic trade (see above). There is little trade in short-shelf-life products, such as bread products, by far the largest proportion of exports being in the form of biscuits (c. £106 million).

The Cereals Sector is probably more heterogeneous than most other major sectors of the food and drinks industry in terms of raw materials used, processing technologies and products. Bread is by far the most important product type in volume terms, accounting for some 52% of total cereal consumption, with biscuits (10%), breakfast cereals (7·7%), domestic flour (6·8%) and cakes (5·7%) being the next four major product types. A wide range of other cereal products includes cake and pudding mixes, pizzas, pasta, infant foods, rice, sago, tapioca, etc., individually accounting for only minor proportions of total cereal consumption. In value terms bread is still the dominant product (36%), but biscuits (20%), cakes (16%) and breakfast cereals (13%) assume more importance because of greater added value.

CEREAL SUPPLY

The production of cereals has increased dramatically in the UK over the last 40 years. The total UK cereal crop averaged 23·3 M tonnes in the five-year period 1982–6 compared with an average of only 7·5 M tonnes over the period 1947–51. Within this overall picture there have been markedly different trends for the different major indigenous cereals. In the late 1940s oats were the major cereal crop in the UK, production averaging 2·8 M tonnes per annum and accounting for over 37% of the total cereal crop. By the early 1980s oat production had declined to only just over 0·5 M tonnes and accounted for only 2·3% of the total cereal crop. In stark contrast both wheat and barley production have grown very considerably. Barley production grew most rapidly in the 1950s and 1960s, but the major increases in wheat production have occurred only in the last 10–15 years.

These changes in production during the last 20 years have been accompanied by marked changes in the structure of the cereals industry at farm level in England and Wales. In the late 1960s there were about 110 000 cereal farms on which a total of 3·18 M ha of cereals was grown. The average size of cereal enterprises at that time was 28·9 ha. By 1986 there were just over 68 000 cereal farms growing 3·42 M ha of cereals, each enterprise averaging over 50 ha. The improved efficiency in production has been due both to improved agronomic practice, and the development of higher yielding varieties, especially those that can respond well to high input systems of farming.

The result of these changes in production technology has been that the UK has become largely self-sufficient in the major indigenously grown cereals. However, some 0·65 M tonnes of non-EEC (mainly Canadian) wheat, at a cost including import levies of over £150 million, are still imported for bread flours requiring high protein level (e.g. wholemeal flours). Furthermore, in years when poor weather conditions at harvest time cause premature sprouting of grain on the ear, much of the home grown wheat crop is unsuitable for breadmaking and other end uses, which necessitates imports of sound wheat. The 1985 and 1987 harvests were both examples of such poor harvests, and in 1985 it is estimated that the cost in terms of the external balance of trade was over £125 million.

The only other major import of cereals is of maize, most of which (almost 0·9 M tonnes worth about £140 million) is used for processing into starch, glucose, germ and oil, plus maize gluten as a by-product that is used for animal feed. About 140 000 tonnes are used in distilling, whilst another 185 000 tonnes are used for breakfast cereals and extrusion cooked products. Some wheat is also used for these purposes, but the starch processing capabilities and extrusion cooking performance of wheat are generally considered to be inferior to those of maize.

MARKET FACTORS

Some general trends that have affected food production and consumption patterns in recent years have already been noted in the Dairy Products and Meat reviews. Bread consumption has declined steadily for several decades, as in many other western countries, probably as a result of this relatively cheap staple foodstuff being replaced by other more expensive and varied items, as the general level of wealth increased.

Cereals have benefited considerably in recent years from publicity surrounding the diet/health debate, particularly with respect to the role of dietary fibre, but also with respect to the perceived adverse effects of high fat and sugar intakes. Cereals are naturally high in dietary fibre and low in fat and sugar, and COMA recommendations, echoed by reports and advice from various other bodies including NACNE, have highlighted the need to increase cereal consumption and to decrease consumption of foods containing high levels of fat and sugar. As a result, there has been a greater demand for dietary fibre-rich products, and this has stimulated considerable new product development.

The breakfast cereals market has profited in particular and the volume purchased rose by 25% between 1979 and 1985 with particularly strong growth in bran-enriched and muesli-type products. The bread market has also been affected, and between 1983 and 1985 the slow decline in bread sales appeared to have been halted. Within these overall bread consumption figures, there have also been quite marked changes in individual bread types. For many years prior to the mid-1970s, white bread accounted for over 90% of all bread consumed, but the proportion has declined for the past 15 years such that it accounted for under 54% in 1986. Brown bread consumption increased from about

5% of total consumption to about 12% between the mid-1970s and early 1980s, whereas that of wholemeal remained at about 2–3% until 1979, but had risen to about 15% by early 1987. Multi-grain breads are also now taking a significant share of the bread market.

Despite the response of consumers to dietary advice in the bread and breakfast cereal areas, there are also contrary trends. Thus, expenditure on biscuits and cake, which are generally relatively high in fat and sugar, has increased during the early 1980s, although it should be said that these foods account for only small proportions of average total fat and sugar intakes.

There has also been mounting pressure recently from various lobby groups aimed at reducing the number of additives in foods and the levels of pesticide and other residues. The Cereals Sector, particularly bread, has come under scrutiny with respect to preservatives, bread improvers and bleaching agents that potentially may be used. The result has been that in the last two years some bread manufacturers have omitted some additives from their products. This has created a requirement for R & D to devise alternative methods of bringing about the technological effects that the use of these additives are designed to achieve, a requirement that was also identified recently in the *Report of the Food Composition and Processing Research Consultative Committee (The 'Gorsuch' Report)*.

CEREALS PROCESSING

Milling

The biggest change in UK milling in recent years has been the change in the grist (blend of different types of wheat) for bread flours. The 'average' bread grist in the early 1960s might have contained 75% of high-protein Canadian wheat, 10% of Australian or Argentinian origin, and only 15% of UK-grown relatively low-protein 'filler' wheat. Nowadays the position is quite different. In 1985/86 Canadian wheat accounted for only about 23% of the 'average' UK bread grist with a further decline to below 20% in 1986/87. The remainder of the bread grist could all be of UK origin, and, except when the crop is of poor quality due to bad weather at harvest time, it usually is.

Undoubtedly, during the last 15 years or so the main driving force behind the decline in the use of non-EEC wheat for breadmaking has been the large differential in price between UK wheat and non-EEC wheat resulting from the imposition of a swingeing levy on wheat imported into the EEC. However, the decline in non-EEC wheat usage had started even before the UK joined the EEC, and reductions in the use of non-EEC wheat and its replacement by UK wheat have been facilitated by several technological developments arising from R & D.

Of major consequence was the development in the early 1960s of the Chorleywood Bread Process (CBP) at the British Baking Industries Research Association (since amalgamated into the Flour Milling and Baking Research Association, FMBRA). It was recognised early in the development of the CBP that white bread of appropriate quality could be produced from flours with protein contents about 1–1·5% lower than those required for the traditional long fermentation process. This realisation, plus the fact that the CBP was adopted by much of the breadmaking industry during the 1960s, soon resulted in protein levels for white bread flours being reduced with a consequently reduced necessity for such high levels of high-protein Canadian wheat.

During the last five to seven years two other developments have accelerated the demise of Canadian wheat. Firstly, in the early 1980s the price differential between UK and Canadian wheat grew to such an extent that it became more economical to bolster the protein content of flours milled from UK wheat by the addition of the isolated wheat protein concentrate, gluten (which is produced commercially), rather than by the use of high-protein Canadian wheat. This had a dual advantage of not only increasing the amount of UK wheat required for milling into flours to which gluten would be added, but also of increasing the demand for UK wheat for processing into gluten and starch. Research has shown that the use of gluten-supplemented flours is also most effective in the CBP. The second factor is the development at the FMBRA of the use of fungal α-amylase as a means of improving loaf volume and crumb structure. This technology only works with rapid breadmaking processes, such as the CBP, and its introduction resulted in the ability to produce bread of an appropriate commercial standard from yet lower protein content flours.

As the consumption of white bread has decreased and that of brown and wholemeal bread has increased, the patterns of flour production have changed in step. To produce wholemeal bread of a quality acceptable to the consumer requires flour of much higher protein content than that required in flours for white bread; this is due to rather ill-defined adverse effects of bran and germ fragments on baking quality. The major use of the high-protein Canadian wheat that is still imported into the UK is, in fact, in the production of wholemeal flours.

Because UK millers are now so dependent on the UK wheat crop, its quality and consistency of quality have become of prime concern. There is a large number of breadwheat varieties that farmers may grow, but there are no regulations as to whether or not a wheat variety must have better milling or breadmaking quality specifically before that variety is allowed to be grown. Variations in climate and soil type can be quite significant even within one farm, let alone between districts, and, since wheat is often delivered directly from farm to flour mill in 30–40 tonne lots, little blending occurs that might otherwise even out quality variations. This places a large burden upon the miller for testing the quality of individual wheat deliveries at the mill gate, and R & D has provided methods for making such evaluations, ranging from estimation of protein and moisture content through to analysis of varieties present in wheat parcels by electrophoretic 'fingerprinting' techniques.

Finally, it should be borne in mind that there are links not only vertically in the food chain from producer to processor to consumer, but also laterally. The imposition of production quotas on the dairy industry is a case in point. The milling of white flour generates by-products, such as bran and 'wheatfeed', the latter being rich in germ, but in a rather finely fragmented form. Some of the bran is used as an ingredient in bran-enriched foods and of late the market for this has increased. Wheatfeed, however, is used relatively little in foods, and most of it is sold for animal feed use. The imposition of production quotas on dairy farmers resulted in a sudden, and apparently unforeseen reduction in demand for wheatfeed, which affected the profitability of the milling industry adversely as prices plummeted.

Baking

The baking industry in the UK also represents a substantial part of the UK food industry, the value of sales being estimated at over £3 milliard. The complexity of the industry is considerably greater than that of the milling industry, with bakeries ranging in size from the small local high-street bakery to large automated high-output plant bakeries. It is also complex in terms of the wide range of products manufactured, which includes numerous bread types, cakes, biscuits, pies, etc. The processing technologies and raw materials used for these various products are rather different, and, as far as the larger manufacturers are concerned, individual bakeries tend to be devoted to a specific product type.

Bread

As noted earlier, bread is by far the most important cereal product type notwithstanding the slow decline in consumption that has occurred over many years. The development and rapid adoption of the Chorleywood Bread Process by the UK industry 25 or so years ago has had far-reaching consequences in other parts of the food chain as far back as growers and breeders (see above). In this process, mixing is carried out at high speed and to a specific work input level so as to develop a continuous gluten protein network within the dough capable of holding the gases formed during fermentation. In traditional long fermentation processes this development of the gluten was achieved by fermenting the dough in bulk for prolonged periods before dividing and moulding and then proofing prior to baking. The intense mixing replaced the lengthy fermentation stage, although proof was still required.

The new process had a number of advantages which were quickly recognised, and, within about 10 years of starting the R & D work, most of the bread produced in the UK was made by the CBP. The ability of the CBP to utilise gluten-supplemented flours and the ability to use fungal α-amylase as a bread improver in this process have further enhanced its value to the UK in recent years (see above).

The CBP was developed for white bread production, which in the early 1960s accounted for over 90% of all bread consumed. However, the process can also be used successfully for wholemeal bread, and this has been important as consumers have switched from white to wholemeal and brown and other high-fibre breads. R & D has led to the introduction of measures for improving wholemeal bread quality, such as the use of ascorbic acid, fungal α-amylase and emulsifiers. Application of such technology has resulted in wholemeal breads being made available with volume and textural properties more akin to those of standard white bread than to those of the traditional wholemeal product. The acceptance by consumers of the message to eat breads containing higher fibre levels, especially wholemeal bread, is thought to have been due in considerable part to the availability of such light-textured breads with improved eating quality.

Consumerism and media pressure have brought other reactions from bakers, who are now producing white bread from unbleached flour, and bread free from permitted preservatives such as calcium propionate. This pressure has been reinforced by the major food retailing chains, which are able to exert considerable influence. Reacting to these pressures may create problems and opportunities for R & D. Omission of preservatives, for example, may create problems for bakers (and also consumers), particularly in warm weather. Alternative preservation techniques would be useful.

Biscuits

The biscuit industry is a compact industry, in which there are relatively few manufacturers and the workforce is relatively small. A considerable number of different lines are produced, although there are, in fact, only four or five major biscuit types, much of the proliferation of lines being due to different secondary processing, such as chocolate enrobement and use of different fillings, as well as some variation in formulation of the biscuit itself.

Biscuit-making processes, which comprise the three main steps of dough mixing, formation of the dough into the required shape, and baking, are highly mechanised, yet, despite this, product quality is governed to a significant extent by the skill of the operators. This is because, unlike the situation with the Chorleywood Bread Process, where control over the dough's processability is effected through control over work input, no such parameter has been identified for controlling biscuit dough mixing. No major changes in the principles of biscuit production have occurred in recent years, but improvements in the various unit processes, such as dough mixing, sheeting/moulding, and baking, have all contributed to improving control over product quality and processing efficiency.

The properties required in biscuit flours are generally almost the opposite of those required in bread flours. For biscuits, soft milling character, low-protein level and protein with weak extensible gluten properties are required. Wheats with such properties are easily grown in the UK, and, indeed, UK-grown wheat has traditionally been regarded as being almost ideally suited to this enduse. There are some suggestions, however, that biscuit wheats that have been produced by breeders in recent years do not have as high a biscuit-making quality as those available 20 and more years ago. These statements would appear worthy of evaluation.

To overcome variation in flour quality for certain types of biscuit, specifically to reduce the elastic recoil of the dough after shaping, which affects the dimensional properties of the biscuit, manufacturers add sodium metabisulphite (SMS) at levels up to 200 ppm. In the present climate of antipathy towards such additives, R & D aimed at removing this requirement would appear to be merited, and, since other countries in Europe disallow SMS, a solution to this problem might also enhance export opportunities.

Flour confectionery

The flour confectionery (cakes and pastries) section of the baking industry produces a very diverse range of products, and, despite an estimated almost two-thirds of all cakes being produced in the home, retail sales of flour confectionery items are still large at over £700 million per annum. About two-thirds of the cakes produced outside the home are made in plant bakeries, the remainder being produced in small family-style bakeries. It is only in the flour confectionery area where any significant level of distribution of cereal products under conditions other than ambient is required, this being due to the perishable nature of fillings such as cream and meat. The market for cream cakes has, in fact, grown rapidly in the last 20 years, because of the development of chill and frozen distribution systems.

As with biscuits, there have been relatively few, if any, very major changes in recent times in the way flour confectionery items are made, although improvements in the various unit operations involved and greater mechanisation and automation have given substantial benefits in terms of efficiency and consistency of product quality.

Again, as with biscuits, the major raw material, flour, is from soft wheat. With cakes, however, flours are treated with chlorine gas both to bleach the flour and to enhance baking performance, particularly in formulations containing high levels of sugar and liquid, which gives cakes with light, fine texture and moist eating quality. With the development of increased consumer interest in additives, attempts have been made to devise alternative means of producing the improving effects of chlorine, but so far without success.

In both the cake and biscuit areas, other important ingredients are fats and shortenings, sweeteners and emulsifying agents. Better understanding of the production factors that control the functionality of these ingredients (e.g. the crystallisation of fats) and the availability of different types within each of these ingredients classes (e.g. oils and fats from previously unused sources such as oilseed rape, palm and sunflower; sweeteners such as glucose syrups, including high maltose or fructose syrups, and non-nutritive sweeteners, such as aspartame) would be beneficial.

An important development in recent years in the cake area, which applies equally to bread, is the introduction of in-store bakeries and similar outlets such as hot bread shops. It is estimated that there are now approaching 1000 such outlets in the UK, and they are estimated to account for about 5% of total bread sales. The products sold in such outlets have considerable consumer appeal, since they are freshly baked, and the development of this form of retailing is having some impact on the packaged cake and wrapped bread markets.

Both biscuits and cakes are targets for reforming nutritionists, because of the high sugar and fat contents of many product types, although consumers do not appear to have responded exactly wholeheartedly to messages to avoid them to judge by the sustained spending levels on these products. Manufacturers of both product types have, however, introduced new products with a healthier image to attract the health-conscious consumer. Such products range from bran-enriched cakes and biscuits to muesli biscuits and various cereal bars.

Breakfast Cereals and Other Products

The breakfast cereals area is the remaining large section in cereal food products with sales value of about £550 million per annum. The industry comprises only a handful of large producers, manufacturing a relatively limited range of products. The industry has enjoyed a buoyant period in recent years, benefiting considerably from nutritional advice to increase dietary fibre intakes. Muesli sales have grown rapidly, as have sales of other ready-to-eat cereals with a healthy image, and a number of new products have been introduced, particularly bran-enriched or whole grain-containing products. Traditional products, especially porridge, have suffered declining popularity for a long time, as consumers have turned increasingly to convenience foods.

Manufacturing processes for breakfast cereals tend to be specific to individual products and mainly involve some form of steeping, forming, heat treatment and drying or some form of puffing. Extrusion cooking has started to replace traditional processes (e.g. replacement of steeping, forming and steam retorting in the manufacture of cornflakes), because of lower capital costs, lower energy requirements and flexibility of the process. However, consumers generally find that the eating characteristics of extrusion-cooked products cannot match those of the traditional products, and this is limiting the uptake of this relatively new processing technology. Breakfast cereal manufacture accounts for a substantial proportion of the cereals imported into the UK, especially maize, but raw-material type again is largely product specific.

Extrusion cooking is also used widely in snack-food manufacture, where novelty products seem to be all important. Raw materials used are potato granules, maize and to a lesser extent wheat flour. The retail sales value is estimated at about £120 million per annum. Other major products that are produced by extrusion cooking include flat crispbreads, but, as with breakfast cereals, the promise of the new processing technology does not seem to have been fully realised as yet.

Starch and Gluten Production

A use of wheat that has expanded considerably in the past seven or eight years and is still expanding is in the production of gluten, starch and glucose syrups of various kinds. This industry now accounts for a significant proportion of wheat milled into flour (about 5%, equivalent to about 0·24 M tonnes per annum). The development of this industry has been stimulated by the demand for gluten in breadmaking flours. The starch co-product is sold as starch to the food and other industries (the paper-making industry in particular), and it is also processed into glucose and glucose syrup for use in food manufacture, production of potable alcohol and other fermentation applications, and as a starting material for some chemical production e.g. citric acid. Most of the requirements for starch and its derivatives were met 10 years ago from imported maize, but home-grown wheat now accounts for a substantial proportion, and, indeed, manufacturers who previously used only maize, now also have adjacent facilities for wheat processing.

STRATEGIC OBJECTIVES OF R & D IN THE CEREALS SECTOR

The general objectives of the Cereals Sector are not unlike those in other sectors and include:

— Use of high standards of manufacturing practice in converting cereal grains into a wide range of attractive, nutritious and wholesome products, from which the consumer can choose, and into ingredients and raw materials to be used in the food and other industries.
— Maximising the use of UK-grown cereals, elimination of cereal raw-material imports, and enhancement of cereal-grain exports.

— Development of lowest cost methods of manufacturing cereal-based foods, whilst at the same time producing traditional and new products of the highest possible quality.
— Development of new technologies for utilising cereal grains in food and non-food applications, development of new markets (including export markets) and expanding existing ones, and transfer of technology both from and to other sectors of industry including the food, drink, chemical and electronics industries.

RESEARCH AND DEVELOPMENT WITHIN THE CEREALS SECTOR

No official statistics are available for industry funded R & D expenditure within the Cereals Sector, but it is likely that this is of the order of £10 million per annum. As would be expected, most of this is near-market research and product and process development, but some relatively basic research is also supported in universities, research institutes and research associations as well as in individual companies. Most of the large milling, baking and other cereal products manufacturing companies have their own R & D facilities. Many of these large companies have extensive food and other manufacturing interests outside the Cereals Sector, and an individual R & D facility usually carries out work covering all the parent company's interests.

Most companies in the Cereals Sector are members of one or more of the research associations (RAs) in the food area, where both pre-competitive collaborative research and confidential, single company, contract research is carried out. The milling and baking and associated industries operate their own RA, the Flour Milling and Baking Research Association (FMBRA), which was formed on amalgamation of two predecessors, the Research Association of British Flour Millers (RABFM), which was established in 1923, and the British Baking Industries Research Association (BBIRA), established in 1946. The RABFM was, in fact, one of the first RAs to come into existence after the inception of RA schemes at the end of the First World War. The FMBRA is approximately equally supported by industry and MAFF, and much of its work is concerned with improvements in processing technology, raw material evaluation and selection, the nutritional quality of cereal-based foods, prevention of microbiological spoilage, and basic research on the composition of cereal grains and changes that occur during processing and subsequent storage.

Both cereal producers and users also pay levies to the Home-Grown Cereals Authority, and part of this money is to support R & D relevant to their interests. Government funds much of the R & D on crop production, although many plant breeding companies have their own research facilities, and research on crop protection, fertilisers, etc. is carried out by the chemical industry. Unlike the dairy, meat and vegetable sectors, there has never been a government research institute specifically devoted to cereals, although the AFRC Institute of Food Research, Norwich Laboratory, now does a significant amount of work in this area.

REQUIREMENTS OF THE CEREALS SECTOR FOR PUBLICLY FUNDED RESEARCH AND DEVELOPMENT

The Cereals Sector, in common with other sectors of the food industry, perceives its need for publicly funded R & D as being for underpinning research, whether this be basic or applied in nature. The industry sees its own responsibility as being to carry out applied R & D that has clearly defined commercial objectives. Adaptation of the results of underpinning R & D to industrial needs, or technology transfer, is seen as being a shared responsibility. Areas of science and engineering that are important include the following:

Raw Material Composition and Changes During Processing

Raw materials represent the cereals industry's major cost; in milling, for example, they account for 80% of costs. The present very high level of dependence on UK-grown cereals presents considerable problems because of the variability in the UK crop. The problems this creates include difficulty in producing intermediate products (e.g. flour) of consistent quality and difficulty in controlling processes to give consistent end-product quality for the consumer.

The changes that individual components undergo during processing, and the interactions between components, which processing brings about, are imperfectly understood. This again creates difficulties in devising methods for controlling processes and hence controlling product quality, and in some cases it limits the replacement of imported grain by UK-produced grain.

It follows, therefore, that greater understanding of raw material composition, of the factors that affect quality, and of changes that occur during processing are major requirements. Composition should be thought of in a broad sense: it includes not just a description of the chemical components present but also of their structures at different levels of organisation and their physical behaviour. This knowledge will lead to breeding of better cereal varieties, development of better raw-material specifications and of better methods for evaluating processing quality, better control over manufacturing processes and better quality in products sold to the consumer. It may also allow products to be made with minimal use of some of the additives that are now needed to make products of acceptable quality.

Research in the UK into the structure of cereal grains and their subcomponents, and into the factors that control cereal-grain quality, is second to none, and continued research funding is essential in this area to support the efforts of plant breeders, cereal growers and grain users to make maximum use of the UK cereal crop.

Processing Operations

There is continuing need to understand the physics and engineering principles behind the unit processes involved in producing cereal-based foods. This is necessary for improving manufacturing efficiency, improving process and product-quality control (including microbiological quality), and also development of novel structures and textures in cereal foods. Research in this area would benefit both food producers and machinery manufacturers.

Since most cereal food manufacturing processes involve the use of dry ingredients (e.g. flour, semolina, grists, etc.) as starting materials, particulate technology including the physics of operations, such as milling and sieving and the mass flow behaviour of dry particles, is an important area. Most processes also involve mixing ingredients usually as wet systems to obtain an intermediate mix of appropriate consistency. The ability to predict and achieve a constant consistency is important for controlling the process and the quality of the product. Production of most foods in this area also involves a heating step as part of the process of creating the product's structure. More needs to be understood about the physics of heat transfer (both heating and cooling), again to improve processing efficiency and to improve control over processing.

Process, Automation, Process Control and Sensors

To minimise costs, cereal foods are increasingly made in relatively few, highly automated, large production facilities at high throughputs, and with low manning levels. This necessitates sophisticated control systems to ensure efficient processing, consistent product quality and minimum waste. These control systems are dependent on appropriate sensor techniques and the ability to respond appropriately to the signals produced. The main need here is not for work on the development of the electronic side of the control systems, since these are common to other sectors of the food industry and other industries, especially the chemical industry, and are usually already available. Rather the need is for development of specific sensors that will detect the key properties of intermediates and processing conditions and for knowledge of how a change in a property or condition will affect the process and the product (i.e. for mathematical models of processes). This, of course, requires a detailed knowledge of composition and interactions of the ingredients.

Many of the processes in this sector use dry ingredients and involve semi-solid intermediates such as dough. Consequently, biosensors probably do not have a great role to play in this sector, but non-invasive sensors for on-line component analysis (e.g. by NIR or NMR), mass flow, rheology, heat flux, humidity, colour, size and shape measurement are needed. Application of 'expert systems' is also likely to be of value in process control and in other areas such as formulation changes/improvement. The need is to demonstrate the usefulness of existing 'shells' rather than to carry out research on the design of expert system shells themselves.

Storage, Distribution and Maintenance of Product Quality

The Cereals Sector has highly developed distribution systems, particularly for short shelf-life items such as bread. However, with production being concentrated in large production facilities and products having to be transported over wide areas, it is important that product freshness and microbiological quality should be maintained as long as possible. Development of measures to extend shelf life would ensure the consumer of the quality desired in a product, but would also reduce distribution costs.

One of the major unsolved problems in this area is the staling of baked products. Progress has been made in understanding the mechanism, but further research is needed to gain a more precise understanding and to develop methods of preventing it or reducing its rate. Microbiological spoilage of cereal foods is also an area that needs more research to provide the consumer with safe products, of appropriate keeping quality, especially in view of the current pressure to remove preservatives from ingredients lists. In common with most of the food industry, rapid methods for assessment of the microbiological quality of ingredients and products is an important requirement.

Nutrition and Food Safety

Cereals now have a high profile because of the dietary fibre issue. However, there are a number of questions still to be answered about the physiological role and long-term health benefits of components of the dietary fibre complex. The nutritional interactions of these different components also remain to be defined, as do their interactions with other dietary components such as minerals. Cereals are rich sources of starch, and it is now thought that starch has a much wider physiological role than simply being a source of calories. Some starch (resistant starch) appears to have the physiological action of certain types of dietary fibre, and there may be other effects also. The roles of starch and dietary fibre need to be defined more precisely so that better nutritional advice can be given both to benefit the consumer and to reduce the burden on medical services.

Certain baked products are high in fat and/or sugar and, although for most consumers such products form a relatively small proportion of total calorie intake, research is nevertheless needed to find ways of reducing these levels and increasing the P/S ratio in the fats used, yet at the same time maintaining the visual, textural and organoleptic properties of the traditional product. Decreasing additive use is a requirement common to most food sectors, and the Cereals Sector is no exception. This is likely to come about through better understanding of the composition of ingredients, intermediates and products, and better understanding of unit processes.

Quality assurance in terms of agrochemical residues and mycotoxin contamination is an important consumer concern. More rapid and cheaper methods of monitoring for such contaminants are required for screening raw materials. This is particularly important now with the trend towards whole grain products, since the contaminants, if present, tend to be concentrated in the bran and germ. Mycotoxin levels, in particular, need to be kept under surveillance, especially in view of the developing trend towards 'organically-grown' grain as the raw material source.

The effect of processing on the nutritional value of cereal foods requires study, particularly for foods produced by newer processes such as extrusion cooking. The possibility of cereal products producing food intolerance, apart from gluten enteropathy, although only a minor consideration, needs to be viewed in the light of food intolerance in general.

CONCLUSIONS

The Cereals Sector is an important and complex part of the UK food industry. Cereal-based food products make a major contribution to the UK diet and they have achieved a much higher profile and more favoured position as a result of recent dietary recommendations.

There have been major changes in the cereals supply situation over the last 20 years or so, which have been facilitated by new developments in processing technology, particularly in breadmaking. The current reliance on UK-grown wheat as the industry's major raw material has created problems for the industry, because of the variability in quality, and quality assessment of raw materials has assumed great importance.

The Cereals Sector invests considerably in R & D, the emphasis of this work being on research with clearly defined commercial objectives. The milling and baking industries run their own research association, and both cereal producers and users support R & D through levies paid to the Home-Grown Cereals Authority. The industry looks to the public sector to fund both basic and applied research that underpins its own efforts to improve efficiency, develop new products and processing technology, and to provide the consumer with safe and nutritious cereal-based foods. Examples of successful developments arising from the partnership between the public and private sectors in supporting R & D are as follows:

— Development of the Chorleywood Bread Process.
— Development of fungal α-amylase as a bread improver.
— Use of near infra-red reflectance methodology for off-line and on-line monitoring of wheat and flour composition.
— Replacement of imported wheats by home-grown wheat for use in breadmaking through provision of appropriate wheat varieties, improvements in agronomic practices giving grain of appropriate quality and changes in processing technology.

There are still considerable opportunities to increase the utilisation of the UK cereals crop, to increase the efficiency of the cereals processing industry and to provide consumers with improved or novel product concepts. Continued public sector investment in underpinning R & D to support those efforts is essential if these objectives are to be achieved.

Report of Discussion

Rapporteurs: M. G. LINDLEY

Tate & Lyle plc, Reading, UK

and

D. MCHALE

Cadbury Schweppes plc, Reading, UK

While the theme of the Conference was the Food Chain, and the papers in this session dwelt on the vertical strands of their own particular food sector chains, much of the discussion on this session focussed on the concept of a food net. Dr Crossett and Dr Davies developed the theme in opening the discussion. They stressed that the vertical strands of the individual food chains form a network embodied in the wider economic and social net, and that the ramifications of the network must allow consumer needs to be met economically and effectively.

The concept of the network has implications for food research priorities, but priority setting probably becomes more difficult because of the need to integrate the chains. There seemed to be some measure of agreement that to seek a more integrated management of the food chains was desirable, so easing the problem of developing concensus on research priorities, but communication channels were not yet in place to do this successfully. Therefore, the key issues to address are:

(i)   the development of appropriate lateral links between chains;
(ii) the mechanisms by which views may be exchanged effectively; and
(iii) the identification of the real factors that influence the priorities.

While this concept of the food network with its lateral connections between chains is important, in contrast, by focussing on 'the chain', there is the advantage that research can be more purposeful, with a clear direction from the producer to the consumer. It is important to recognise the impact research on any specific link in the chain can have on the chain as a whole. Isolated research without this appreciation will rarely be exploited. To ensure that the end products of the food chain are of maximum quality, it is critical that the raw material variety, maturity, and harvesting and production methods are considered fully. As an example, the rapid growth in the development of chilled foods affects the agricultural methods used, emphasising the need for a greater understanding of the influence of time–temperature relationships on product quality. This particular development also illustrates the vital role of consumer education in relation to innovation, nowhere more so than in relation to irradiation. It is important to recognise that the consumer is not always adequately informed and that this lack of knowledge contributes to a reluctance to accept the scientific, or expert, view. Should scientists merely inform, or should they also interpret and persuade? Clearly, the scientific community bears a heavy responsibility for risk analysis and must respect the views and fears of consumers. For scientists to retain credibility, the facts must not be presented in a way that destroys the trust, for the consumer needs to see and be convinced that the benefits from change outweigh the risks. The danger that, whatever the priorities, the value of successful research is destroyed by inadequate communication must be avoided.

These threads of discussion raised few points of conflict with the participants. There was a measure of agreement, but the discussion to this point had considered merely generalities that were difficult to fault. The question of 'priorities' had not been addressed, nor had the critical question of ensuring an appropriate balance between public and private funding. Whereas government is encouraging universities and research institutes to adopt a more applied approach to food and agriculture research, in general, food manufacturers would prefer to see the academic community concentrating their resources on more long-term, fundamental research. Industry is, in general, not in a position to carry out fundamental research, for its priorities change too rapidly and it cannot easily justify the long-term commitment necessary. However, the long-term research base is fragile and must not be eroded. To ensure its survival, universities and research institutes must be free to commit significant resources to basic research. Clearly, such basic research must be monitored carefully. Better accountability is required, the research must be focussed and its results must be communicated effectively. However, while industry maintains that the funding for such work is the province of government, government expects to concentrate its efforts on those areas where clear benefits can be visualised in advance of carrying out the research. Clearly, there remains an area of conflict here.

To conclude, the wide-ranging and somewhat unstructured discussion was almost inevitable, though its inevitability did not allay some measure of disappointment. 'Priorities' as an issue were not addressed. It would have been satisfying to report a consensus. However, agreement on specific priorities for research was not forthcoming, except within the individual commodity areas, nor was there a meeting of the minds on how fundamental food and agriculture research should be funded. What is clear, however, is that there is the danger of having an inadequate fundamental research base to apply to the food and agriculture industries. Should that happen, the implications for these industries are serious indeed.

THEME 4

Implications for Policy

Introduction*

J. S. MARSH

Department of Agricultural Economics and Management, University of Reading, UK

The purpose of this theme is to identify ways in which an approach which takes account of the Food Chain as a whole, illuminates analysis of *policy issues*. Much policy analysis is devoted to the implications of specific policies for separate segments of the food chain. Agricultural economists have studied the implications of policies for farm management, for the consumer and for the economy as a whole. Political scientists have examined the interaction between pressure groups and political systems in the formation of agricultural policy. The analysis of the interaction between food processing firms and food distributors has interested industrial economists, especially those concerned with competition policy. Economic analysis and institutional behaviour has figured in work on agricultural marketing. At the retail level sociology and psychology, too, have played an important role.

Such examples draw attention to the wide range of skills already deployed in the analysis of policy related to the food chain. However, each element of the food chain is affected by the performance of all the other parts. Changes in consumer attitudes, say a preference for 'organic food', affect not only the retailer but also the processor, the farmer and firms who supply farm inputs. The nature of these interactions is complex, but for policy to achieve its objectives they have to be taken into account. For example a policy which seeks to sustain prices and limit production by the application of a quota may initially raise the incomes of quota recipients — as intended, but it may also reduce

*Editorial assistance with this theme is acknowledged from Rupert Loader and James Burns.

incomes of processors who find themselves burdened with redundant plant and of consumers whose cost of living is increased.

The contributions to this theme therefore seek to explore ways in which the analysis of the whole food chain aids policy formation and appraisal. A key function is to tease out some of the more important relationships in such an approach. To do so involves inputs from industry, from policy makers as well as from academic research workers.

ECONOMIC INTERDEPENDENCIES

Much agricultural policy exists because politicians, and those who support them, do not like the expected economic outcomes of a market clearing system. Amongst the features which are often rejected are the resulting income distribution, the balance between domestic and imported supplies and the volatility in prices for both farmers and consumers. This rejection of a free market has led to a variety of policy instruments. The way these have been implemented in the European Community is outlined in the paper by Mr McClumpha. Inevitably where the effect is to protect farmers, this leads to production in excess of the amount which would occur in a 'free market'. The international trade repercussions of this have now become an urgent issue for the Community's politicians. As Mr McClumpha's paper makes clear they have for a long time had serious implications for its food processors and exporters.

Sustained imbalances between production and consumption have profound consequences for the whole of the food chain. They mean growth for some types of domestic businesses. More farm inputs will be produced and sold. First stage processing and the storage and transport of agricultural products will be on a larger scale. At the ports special export installations may be needed. Other domestic industries, based on imports, may be forced to contract, whilst all users of agricultural inputs will need more capital to finance their enterprise. In some other countries producers may be forced to cut production. In contrast, if the level of protection is unequal between commodities, other overseas suppliers may discover a new and growing market in substitutes. Thus agricultural policies rapidly spill over into issues which affect firms which form part of the food chain in both Developed and Less Developed Countries. These issues are explored in Mr Harris's paper.

In addition to such issues of the balance of supply and demand the food chain approach is needed in the analysis of the structural characteristics of the industry. These matters are discussed in Professor Thomas's paper on Competition Policy. The interface between the various elements of the industry is affected by the pattern of firm size, the control of capital, the degree of concentration at various levels as well as changes taking place within these structures. Traditionally analysis has assumed there to be numerous farmers, selling via smaller numbers of middlemen, to large processors who in turn sold to a wide variety of retail outlets. Such a model is inappropriate to today's world. At the retail level food distribution has become increasingly concentrated until a large proportion of sales is made by a small number of large supermarket chains. Their sourcing policy powerfully influences the fortunes of many processors and the direction from which raw materials are drawn into consumption. Processors, too, include many large firms, some of which are part of multi-national conglomerates rather than simply operators in the food sector. The co-existence of small and medium sized competitors, as well as foreign suppliers willing to supply branded or own label products at marginal cost represents an important safeguard to competition. At the farm level, too, co-operatives and marketing boards influence the terms of access to some raw materials. In a few areas, notably poultry and pigs, concentration among producers has gone so far that some buyers, for processing or distribution, can choose to deal for a large part of their needs with a small number of named suppliers on a long term contractual basis.

A 'food chain' approach is needed to identify many of these consequences of the changing characteristics of competition in the industry. To concentrate on only one segment raises the risk that important social and economic repercussions may be missed.

POLITICAL INTERDEPENDENCIES

The formation of policy is profoundly influenced by the institutional structure of government. In democracies the ultimate sanction is for a government to be voted out of office and food issues can seriously threaten a government's survival. Scarcity leading to famine is only the most acute example. Sudden surges in food prices, or even in the price of particular foods, can cause political embarrassment and a prime justification for agricultural price support has been to avoid such calamities. However, a complex set of other political considerations influence the way farm policy operates. These include the activities of pressure groups which involve all parts of the food chain.

Within the food and agricultural sector it is easy to identify some homogeneity of interest between those pressure groups representative of production on the farm (Farmers' Unions, Manufacturers of Feed and Fertilisers, Landowners, etc.). Even here important differences do exist; between for example small and large farmers. Processors' and distributors' interests diverge more obviously and tend to be focussed on lower prices and higher quality. While consumers share this concern, more than half the cost of their food is added after the product leaves the farm gate, so for them the activities of processors and distributors is as important as that of farmers. Environmentalists — the recipients of some benefits or costs of farming for which they do not pay or which they cannot avoid — form an increasingly important pressure group and their criticisms influence the way policy is formed, affecting the whole food chain. Dr Grant's paper reveals how in the political debate key issues have come to play a decisive role in the determination of policy particularly in relation to food additives.

Political concerns go beyond domestic considerations. For example, in Europe the reform of agricultural policy is constrained, among other things, by the conflicting interests of member countries and by the widely held desire not to undermine European unity. In the wider international arena, attitudes within GATT on agricultural policy have to take account of political realities within countries as well as the more objectively measured dimensions of economic gain or loss. There are also strong lobbies anxious to encourage development among poorer countries, many of whom are former dependencies of European powers. Mr McClumpha deals with many of these issues in his paper.

This diffuse and shifting pattern of political influence operates through the formation or dissolution of coalitions of interest groups. The 'bandwagon effect' can be of great importance to the food industry. Bans on hormones, legislation relating to labelling, public attitudes to additives and nutrition have all demonstrated the power of such coalitions of interest. In some cases food industry businesses across the whole chain may need to adopt a common stance if their voice is to be heard. For the industry some analysis of the operation of such groups, their tactics and strategy, is required if the development of policy is to be understood.

THE IMPLICATION OF NEW TECHNOLOGY

All parts of the food chain are affected by technological developments. Some originate within the food businesses while others derive from changes outside. In both cases there are 'food chain' effects which need analysis.

Two examples illustrate linkages which have policy implications. New retailing methods with much greater ability to identify quickly which products are 'moving' and which are not, change the competitive position of food processors who face an increased risk that a slow selling product may be 'de-listed'. New methods of producing meat substitutes from vegetable materials may influence the future size of market for livestock farmers and the animal feed industry. Several of these issues figure in Professor Thomas's paper and its discussion of the 'two edged' nature of competition in the food processing sector.

Policy has to concern itself with such issues. It needs to ensure that foods are safe and nutritious. It has to take account of the social implications which may flow as traditional occupations are displaced. In fixing prices it has to take account of the implications of rising productivity for the cost of the policy and the pattern of international trade. In seeking to ensure economic efficiency it has to examine the effect of shifts in competitive power. Since much technology derives from publicly funded research, development and education policy makers have to explore the direction and level of such funding in the context of national needs.

THE MULTI-DISCIPLINARY NATURE OF POLICY ANALYSIS

The significance of new technology, like the need to understand political structures, emphasises that if policies are to be understood, and relevant new policies devised, there must be input from a wide range of disciplines, leading to an interaction of ideas and representations.

The implications for the interaction of scientific work with policy making are important. By their nature many research activities tend to be a fragmentation of effort. Solving difficult scientific problems requires great singlemindedness and concentration of effort. The policy maker, in contrast, has to draw together such elements, to have them interpreted and to relate them to his perception of the goals of society.

By conducting research in an environment which is consciously related to the food chain as a whole, policy implications of research should more rapidly be identified and communicated.

The food chain concept is significant in terms of policy relating to the direction of research. It is clearly not enough to say 'more should go on food' and less on 'production related research'. Such generalised statements have to be broken down into more detailed researchable topics. They also have to be appraised. This, too, is an area in which additional studies are needed if objective assessments are to be realised.

This theme is the result of articulating some of these issues, with a view to encouraging a fruitful dialogue on policy and contributing to the better direction of public funding in relation to the food chain as a whole.

International Trade Implications

A. D. McClumpha

The Nestlé Company Limited, Croydon, Surrey, UK

ABSTRACT

In spite of the importance of value added exports to the economic activity of the Community, the EEC Commission has not been able to recognise this and implement a policy for the disposal of agricultural surpluses at the same time.

Political pressure for high producer prices not related to market demand has created substantial budgetary cost and has led in turn to the discouragement of the importation of agricultural products.

The CAP has been dominated by short term budgetary concerns resulting in:

— *a steady increase in first stage producing and production for sale directly to intervention.*

— *concentration upon bulk volume production rather than the achievement of added value.*

— *pre-occupation with short term policy and action instead of market development and the encouragement of reciprocal trade.*

— *an over developed regulatory system in an attempt to solve short term market distortions and the subordination of international trade to the needs of the CAP.*

Under the Common Agricultural Policy international trade is seen as the means for disposal of agricultural surpluses to allow the achievement of a balance between production stimulated by domestic price, and consumption demand. Over the years lip service has been paid by the Commission to the importance of value-added exports, but administrative decision has never encouraged the aggressive long term export of

Community labour, capital and skills through the products of the food processing industry. Emphasis has been upon short term, high volume disposal of surplus agricultural commodities processed only to the first stage, necessary for preservation.

The political pressures to increase and maintain producer prices at relatively high levels, often unrelated to real demand, have led to growing agricultural surpluses and a high budgetary cost due to the intervention purchases and/or low price export disposals to maintain the agricultural regime. Defence of producer prices not justified by demand has been argued as the need for a 'balanced market', Community self sufficiency and Community preference. The demand of agricultural policy upon the EEC budget has focussed political action upon the cost of export disposals and the short term measures needed to regain equilibrium such as special butter sales to the Soviet Union, rather than upon the longer term fostering of a continuous export business using a full range of community resources.

Similarly, the importation of agricultural raw materials or even processed foods, not adequately produced or not otherwise available in the community, is discouraged by the administration of the price regime, on the assumption that consumer demand will thereby be switched toward community produce. The use of threshold prices, general coefficients and regional levy calculations as well as minimum import prices and the frequent suspension or denial of process inwards relief, substitute administrative systems for market value and fail to recognize quality and variety differences in a wide range of commodities.

The effects of the philosophy and administration of the CAP have been cumulative and by their influence upon business decisions have fundamentally influenced the position of the community food industry in international trade. Present activities and attitudes toward export and import have been shaped by the restrictions imposed upon the industry by the rules of the CAP. In addition, the problems of the CAP itself, especially the need to handle surplus disposals have forced the Commission to think and act in terms of budget protection and short term expediency. The following consequences have developed within the Community

A STEADY INCREASE IN FIRST STAGE PROCESSING AND PRODUCTION FOR SALE INTO INTERVENTION

High agricultural support prices make sale into intervention virtually risk-free in certain commodity areas particularly, butter and skim powder, wine and beef, while the regime operated for sugar and certain fruits and vegetables does nothing to encourage the marketing of surpluses within the community. The production cost coefficients used to calculate the intervention or support price levels are usually generous enough to allow primary processors to organize their businesses to provide adequate margins without the need to provide for the product development or changing consumer demand. Evidence is seen in:

(i) The pressure to set up a scheme for alcohol production from surplus sugar and cereals, which could only exist by means of a substantial subsidy.
(ii) The strict application of quota to iso-glucose manufacture in order to prevent competition for sugar.
(iii) The lack of response to Community efforts to curtail butter and skim powder production until measures became very severe.

A CONCENTRATION UPON VOLUME PRODUCTION RATHER THAN THE ACHIEVEMENT OF ADDED VALUE

With a comprehensive intervention system it has been much more profitable for the farmer to concentrate upon high yields rather than marketable qualities, so long as the commodity concerned can meet intervention standards. The result has been increasing production of feed wheat, medium grain rice and low quality wines as well as butter and skimmed milk powder for which no obvious market has existed. The first stage processing industry has found it necessary to invest in processing facilities to cope with these unmarketable surpluses and agricultural production has therefore been supported by industrial investment simply to process surplus into a storable form.

The part of the food industry closest to the consumer or using CAP materials in high added-value products has often been unable to influence agricultural production toward the optimum qualities that it requires, or to obtain recognition of its role in product development and marketing to the consumer, and has therefore sought to avoid the CAP regime, where possible, particularly in the development of new products. Elsewhere the industry has found it necessary to use sub-optimal raw materials where the price level offered to relieve intervention stocks has been an effective subsidy for their use. These tendencies can be seen:

(i) In the subsidized usage of intervention butter for ice cream and biscuits and cake, where the industry uses butter simply because of price, although butterfat could be used more appropriately in another form.

(ii) In the rapid development of vegetable fat technology during the last 15 years, compared with the lack of similar technological progress with dairy fat.

A CONCERN FOR SHORT TERM POLICY AND ACTION

The dominant pressures in CAP surplus disposal have been those directed to export. Destruction of surpluses has been politically unacceptable and any restriction of intervention or producer price has been unthinkable until recently. World markets have accepted EEC low-priced sugar and the USSR has been the last opportunity for the disposal of heavily subsidized butter. These export disposals have been costly and often made on an *ad hoc* basis, governed by the age and condition of the commodities concerned and the pressure of the EEC budget.

A result of this has been that regular export business of added-value products to meet genuine consumer demand, has been subject to sometimes capricious change in export restitutions interfering with orderly long term market development and promotion. Export restitutions for added-value products are seen as subsidies (e.g. the pasta case[†]) and a cost to the EEC budget in the same way as surplus disposals for export. The long term nature of this type of business and the need for continuity is not necessarily extended from one budgetary period to the next. The ability of the food industry to add value to agricultural products and to extend employment and capital investment within the EEC is not given potential recognition within the administration of the CAP.

AN OVERDEVELOPED REGULATORY SYSTEM

Over the years the CAP regime has added to the series of controls and market management measures used. Along with the problem of short term surplus disposal within close budgetary restraints, has come the

[†]The US complaint against the EEC export restitution system under GATT, referring specifically to pasta exports from Italy.

need to control expenditure to particular objectives without creating new distortions in the market, between member states, or providing new loopholes in existing regulations. Nevertheless, the complications of the CAP regime often allow those who are expert at interpreting regulations or who may be attuned to one-off deals to gain from the system and the lack of transparency and continuity in decision-making. The Commission tends to see all exporters as traders whom it must outwit, rather than as industrialists interested in continuity, added value and employment. Certain parts of the export regime are now so complex that many food industry exporters regard export restitutions as a windfall, impossible to calculate reliably or to use as a marketing pricing factor. A system designed to encourage exporting to world markets is becoming self-defeating, psychologically unattractive to the intended beneficiary, and widely regarded in the world as an unjustifiable subsidy to be challenged in GATT.

These developments within the EEC have also had significant consequences in the rest of the world, particularly due to the action taken by the EEC to control the disposal of agricultural surpluses. Lack of continuity in export policy, whether due to over optimism concerning the control of future production or to budgetary pressures in the short term, has resulted for many years in world markets being treated as an available outlet for surplus, subject only to price incentive.

International trade considerations have been subordinated to the needs of CAP not only in surplus disposal rather than in market building, but also in determining the scope of import controls with the objective of protecting community agricultural production. The results of this policy have been to deny the EEC food industry the ability to secure a strong competitive position in world markets over the long term for the following reasons:

(i) The emphasis on bulk commodity volume sales of an opportunistic nature (butter sales to the USSR are the most obvious example) have undercut regular export business and secured a low quality cheap price image for Community sales.

(ii) The number of bulk export disposals has created retaliatory sales activity from countries such as USA and Australia, with their own surplus disposal problems.

(iii) The increase in EEC production and bulk exports has swamped the residual world market in sugar and dairy products, and increased the gulf between world commodity prices and EEC domestic prices. Third world problems have been exacerbated.

(iv) Low disposal prices for agricultural produce have encouraged overseas buyers to invest in conversion facilities so that they can add value to EEC bulk commodities. For instance, the decline in the world market for consumer packs of evaporated milks has been due to many countries switching to supplies of bulk milk powder for local packing.

(v) Perhaps less obviously, the quality image of EEC produce has declined with the failure to sustain markets for added-value products. It is much easier for an importer to substitute an EEC bulk commodity with that of another origin if the difference is measured by price alone, rather than quality, reputation and presentation.

(vi) EEC policy in the disposal of surplus also encourages the development of state trading. The Commission has sought the power to enter long term contracts, so far unsuccessfully, but current policies encourage overseas governments to anticipate special deals and to focus attention on price alone. There has been noticeable shock, recently expressed by several governments, at the Commission's refusal to extend Russian-style butter prices to other destinations.

(vii) The annual budgetary basis of the CAP makes policy continuity difficult and in spite of a prefixation system the risk entailed in establishing long term export continuity for branded products is high. It is clear that although a pre-financing system exists for many products the complications of operation are such that it is of little value to the exporter. Similarly, the process inwards relief provisions are either set aside or little-used in most commodity areas. The system is supposed to provide for the maintenance of business for EEC factories and employment when EEC raw materials are unavailable or uncompetitive, by allowing temporary import for processing without levy or duty. In practice this function is not fulfilled and there are examples of business being lost to competitors outside the EEC, especially where the capital and labour input would have far surpassed the value of the raw material content.

(viii) The MCA system designed to equalize currency differences between community members also denies the processor of added-value products the facility to cover his own currency risk and to take advantage of favourable currency movements in export markets. Coupled with the rigidities of payment from local intervention boards this has become a distortion of competitivity in world markets.

(ix) World trade is essentially reciprocal and the reluctance of the community to permit agricultural imports, especially of raw materials (except on disadvantageous terms), makes it more difficult to negotiate adequate access for community exports, whether on balance of payments grounds or through tariff barriers. Although the current GATT negotiations may help in this area, there has also been a significant growth of various non-tariff barriers in recent years.

(x) EEC aggressive marketing of surplus materials has added to trade tensions largely because the policy considerations seem to end with the needs of the CAP. The result has been a threat to general trade relationships, particularly with the USA where retaliation and counter retaliation has been proposed by both sides on the products and raw materials of the food industry, not necessarily directly concerned with the dispute. Examples include the inability of the EEC to implement US maize imports into Spain, the misunderstanding of the export restitution system by the USA, as seen in the Italian pasta case, the threat of US retaliation to the EEC ban on meat produced with the aid of hormones and the damage done to trade relations with a number of ASEAN and African countries by the oils and fats tax proposals.

(xi) A particular concern of the food industry lies in the restriction of its ability to develop new products to meet changes in consumer demand and its freedom to develop and utilize new technology. Such restrictions are perhaps understandable if the industry is seen simply as the preserver of the harvest. However, the restrictions on the production of iso-glucose has made the European industry technologically unable to develop products using corn syrups on world markets, while the attempt to restrict recipe uses of oils and fats in competition with dairy products, is seen as an attempt to stem a change in dietary habits and nutritional needs to which the food industry needs to respond.

The CAP has failed so far to harness the marketing and technological development capabilities of the food industry to act in international trade because policy has considered only domestic agriculture. Concentration on producer price and short term objectives, while ignoring the effect upon international markets and relationships, has had the consequences detailed above. Surplus disposal on world markets has had a cumulative effect on international trade which in turn affects the internal market through the narrowing of options available to policy makers, showing the CAP cannot be operated in future without consideration of the wider implications.

FUTURE CONSIDERATIONS

Changes in CAP have begun under the pressure of the budget and further developments are inevitable. The climate of opinion now recognizes the importance of the relationship between supply and the market-place within the community. Talk of social payments and the restriction of intervention rather than untrammelled production is current. What is still less well recognized is that the world market does not provide an indefinite home for surplus and that it is impossible to ignore the effect upon the Community of its own action in the world. Some recognition of this factor is showing in the preliminary position prepared for the GATT negotiations, but there is still a strong tendency to see the main pillars of the CAP as inviolate and not to be questioned.

Points that must be considered in the future determination of the CAP include:

(i) The political and economic importance of maintaining and enhancing exports of community produce in all forms and the responsibility of the Agricultural Directorate for the consequences of its actions in international trade and industrial employment within the Community. Export and surplus disposal are different functions — one to be encouraged and the other to be avoided — but neither is without consequences for agriculture and the community as a whole.

(ii) Positive concern for added-value exports without artificial impediment, but with the administrative measures necessary for EEC industry to obtain a fair competitive position in world markets. It is necessary to consider the optimum time frame for added-value exports separately from that set for agricultural producers. Administration of the CAP should allow the free development of new products and new technology for both home and export demand. Changing demand must ultimately be met by changing agricultural practices, not by the political retention of out-dated production.

(iii) The present administration of detailed trade rules and marketing conditions is decided upon by agricultural management committees responding to political and domestic pressures and operating under procedures such as 'non avis' and decision-making is not transparent. This system does not seem the most appropriate for the proper development of secure, long term export business making a substantial contribution to the community.

(iv) There are no strong political reasons to seek government to government trade although good international trade relations are vital. Existing systems give a tendency toward political deals or other special short term trades. This is not a satisfactory way to satisfy consumer demand or to build a secure export business with a proper concern for added value.

(v) Questions of quality, consumer choice or industrial employment are as important in export trade as in the domestic market. More weight must be given to the overall interests of the Community in providing for agricultural support.

The present discussions in GATT appear to provide a focus for those most concerned with the policies of the EEC, particularly as agricultural policies will form a major part of the negotiation.

Opposition to EEC surplus disposal activities is clearly seen in the reaction of the USA to a loss of its traditional markets. Similarly the Cairns group is concerned that these practices should be modified.

Third world debt concern, allied with low world commodity prices will result in increased criticism of the EEC in particular commodity areas, such as sugar and vegetable oils. In some supply areas the Community has developed a bulk client business, in milk products and to a lesser extent in cereals, which will be terminated if excessive Community production is finally controlled, but may result in customer requests for continuing subsidy. EEC proposals seem likely to include some form of international agreement in various commodity areas, but the record of political co-operation in the solution of world economic problems is not good.

In this climate it is vital that means are found, and made explicit in policy, to ensure long term continuity of international trade by fully utilizing the added-value expertise of EEC industry, much of which has been devoted to avoidance of the CAP restrictions, and recognizing the vital function of exporting to maintain, not only the viability of the CAP, but a satisfactory level of employment and economic activity within the Community.

Competition Policy

R. THOMAS

Clevedon, Avon, UK

ABSTRACT

The stages in the 'food chain' offer sharply contrasted states of competition from the artificial, regulated market of the primary producer, through the combination of very large multinationals and many smaller firms processing, to the powerful retail groups close to the consumer. Competition policy is primarily of concern to the second and third stages. The issues at each are set alongside a changing government policy on intervention, especially in respect of mergers and take-overs, and of pricing policies. This leads to the question of the changing arena of competition — UK or EEC — and the implications of further concentration of power, whether of processors or distributors and to the difficulty of applying a strictly consistent policy across all three.

INTRODUCTION

Competition policy becomes quite a complicated issue when one studies the affairs of the 'food chain'. At one end there is an apparent host of primary producers, keenly competitive with one another and very suspicious of any mutual cooperation, who nevertheless operate within a highly artificial market situation what with the Common Agricultural Policy and major interventionist agencies such as the Milk Marketing Board. At the next stage there are major multinationals processing alongside a range of smaller firms at various stages of evolution; all are seeking to cope with an increasingly turbulent environment which is further complicated by the interventions of conglomerate predators. The products of both agriculture and food processing then find their way to the customer dominated by the very powerful groups that straddle both retailing through supermarkets and sales through independents or other channels via their share in 'cash and carry' activities. At each stage there is a different arena of competition.

This makes the application of a common policy, such as the 'Tebbitt' doctrine that mergers and acquisitions be judged solely on their impact on domestic competition, only superficially attractive. Perceptions of 'orderly marketing' as in so many other industries, have a knack of coinciding with some reduction in its vitality and effectiveness. Yet the international nature of the arena, especially for the major processors, presents real difficulties. These are intensified where the home bases of the protagonists practise differing policies on international access, whether through acquisition or the structure of the distribution systems.

For the UK three established features complicate the situation. The first is the legacy of interventionist agencies in agriculture derived from an earlier approach to farm support and epitomised in such bodies as the Milk Marketing Board. At the processing stage there is the combination of a highly developed and concentrated capability in a number of internationally orientated majors who represent one of the few cases where UK multinationals are among the top ten companies in Europe. Their strategic role is seen as significant but also as being under serious challenge, first from their fellow large scale processors abroad, and second through the challenge to their whole organisational and economic base with changes in consumer taste, preoccupation with freshness or the absence of additives, and the stress on variety. At the retail stage there is an exceptionally well organised set of firms, a handful of which have a massive influence on the market, but who are in intense competition with one another.

Competition policy has therefore been studied not only at each stage in the chain but as between stages. This then raises the question of its application where there is vertical integration/combination between any two stages.

Competition policy has also to be considered in a wider context, that of the potential sources of new competition, and that also involves the possible effects of new entrants whose prime objective is the profit to be gained by restructuring as opposed to subsequent continued operation in any particular industry.

These three stages are outlined below.

STAGE 1 — AGRICULTURE

Within the primary stage competition policy, as such, has limited impact; it is government policy on competition from imports and the continued role of bodies such as MMB that are important. Here there is a curious conflict of views. It is argued by retailers, and in some instances processors also, that the weakness of competition encourages complacency on the part of farmers, this being compounded by the difficulty of organising cooperation among both farmers and initial processors. The result is a failure to take up market opportunities, whether in the UK or abroad, and even a tendency to fail to match foreign competition on such basic features as delivery to the UK market. Overshadowing such matters is however the cost of production of UK supplies, especially where they are denied access to comparable raw material costs through CAP measures or at a natural disadvantage through their own lack of coordinated marketing; the case of pig-meat as between the UK and both Holland and Denmark is typical.

There is little vertical integration by processors backwards into primary production but forward vertical integration occurs in meat and dairy products. The question of competition with imported or alternative home sources does arise, to some degree, but it is more the issue of overall marketing policy as between the direct consumer opposed to the processor that is significant, as in the case of milk. Here we find an interesting instance of competition between farmer-controlled major agencies and powerful distributive organisations having processing interests, such as Unigate.

Is the balance too heavily in favour of the primary producer organisations or is it their marketing policies, indeed their overall strategies, that are in contention? Important interested parties include the trades unions, that have strong representational roles for milk roundsmen, for example, alongside the commercial interests of major distributors who, in turn, are in direct competition with the supermarkets, and therefore involved in the rivalries associated with imported alternatives. Once again the main arena is more subject to import checks and balances, the non-tariff barriers of EEC, than whether the Tebbitt doctrine has any impact. Longer term there is a real power battle between the retailers and both the primary producers and the processors. The application of policy at the retail stage therefore remains of vital interest to the primary stage.

Clearly access to raw materials, and their sources, is of major interest to processors as must be the degree of competition that remains, whatever the source. The specific cases of sugar refining and access to oils and fats illustrate this. The former is clearly a competition policy issue while the latter is essentially a question of farm support and its funding.

STAGE 2 — PROCESSING

Just as processors may favour an arms' length relationship to an NFU-dominated home agriculture so the same situation tends to develop between processors and retailers but with the important distinction that processors may be heavily oriented to branded products for which alternative sources, marketed as retailers' own brands, are readily available, if not within the UK, then elsewhere in Europe. For the processor therefore competition policy is two-edged. On the one hand processors are in favour of open competition. Newcomers with new products may be interesting to watch, and, if successful, they may be acquired as a deliberate policy. On the other hand the ability of retailers to give opportunities to competitors from abroad, or to pursue own-label lines from alternative sources, whether at home or abroad, presents a serious challenge. When the ferocity of the competition among the top half dozen supermarket groups is taken into account the question of access through these organisations, and the terms upon which it is secured, become issues of competition policy on several counts.

The first is the exercise of retailer bargaining power to secure additional discounts on branded lines upon which the Office of Fair Trading has sought to get confidential information before judging whether there was any unfair or restrictive practice. The present presumption that, as long as there is competition that appears to favour the ultimate customer all is well, continues. Indeed in one of the few sectors where there is a form of vertical integration, the tied house system, the brewers now find themselves the subject of yet another Monopolies and Mergers Commission investigation despite the massive development of off-licence sales through supermarkets.

The second is then the quality of opportunity as between UK and foreign processors, more in terms of penetrating each others' home markets through acquisition or direct penetration. Given the extent to which the main UK multinationals already manufacture abroad this is now not a serious problem, but it has been in the past with the appearance of bidders from countries where reciprocity of such opportunity was lacking, at least at the time of particular bids.

This is linked to two further considerations. There is the bid which has to do with competition through specific sector penetration. This was the rationale of the original Elders IXL bid for Allied-Lyons.

In the event this was overtaken by the opportunity for Elders to achieve market entry through purchase of Courages, an opening created by the Hanson take-over of Imperial Group. That take-over illustrated the other consideration, the acquisition of a Group with a view to its total restructuring, with the prime objective being the gains secured through that act and the subsequent successful disposal of a large part of the activities that had been included in the Imperial Group. This, while of a size to warrant, prima facie, review by the Director of the Office of Fair Trading, did not bring about the concentration of control within a sector that had been posed by the various acquisitions by Distillers.

Here there is an interesting aspect of competition policy. Should it have any bearing on the operations of those predator companies that appear to specialise in restructuring activities? These could conceivably have direct implications for competition, as divestment and strategic trades follow from the acquisition of a conglomerate. This has to some extent occurred in both the Hanson/Imperial case — in respect of motorway catering concessions — and the Guinness/Distillers saga. Brewing and milk distribution have demonstrated 'strategic trades', at least on a regional basis.

For the food and drink industries these are very important issues as it is debatable whether the groupings built up on the basis of past market trends are necessarily the most appropriate for the changed situation that is now developing. To the directors of such groups there is the question as to whether they should restructure voluntarily to produce a more relevant organisation — as has been attempted by Cadbury/Schweppes, or wait to be dismembered by a predator. It remains open to debate whether the argument about strategic leadership being dependent on the continued group strengths of UK multinationals is as strong. Much must depend on the value of some countervailing power to that of the retailers. This is a far cry from the days when the boot was on the other foot and the processors enjoyed a very much more powerful position.

There remains the possibility of vertical integration forwards into distribution. The opportunities for this must be slight given the reverse possibilities; the weakness of so many manufacturers in their interface with retailers then leaves two problems. The first is the effect of this on their ability to reduce their very heavy dependence upon the home market, especially if more and more product specification is in the hands of retailers instead of being processor innovation. The point made in other trades is that UK retailers have little direct experience of retailing elsewhere, and of the very special national features, especially in food markets. After all there are very considerable regional differences even within the UK as between say the North-East of England, the Home Counties or Wales. The second lies in the concentration in the process of transport and distribution within the UK and the effect that this may have on the chances for individual processors, especially newer smaller producers.

STAGE 3 — DISTRIBUTION

While the concentration of retail distribution power in a small number of supermarket/foodstore chains may be the main focus of attention, it is important to recognise the share of total food distribution that is accounted for by the catering industry in its widest sense. This covers hotels, restaurants, institutions — both private and public — and industrial 'canteen' provision. These tend to be supplied through wholesalers rather than 'cash and carry' organisations, some of which are in the same groups as key retailers of the supermarket type. There is also the opportunity for direct processor sales to caterers, particularly in the prepared meals category.

Competition among wholesalers and specialist trade suppliers is therefore important, and, while there is some evidence of further concentration here, it has not attracted the attention of that among retailers. Industrial and institutional catering is increasingly provided by contract caterers that are specialist divisions of larger groups, not all of which feature elsewhere in the food chain. Thus P & O own the Sutcliffe organisation. Gardner Merchant, who are prominent in functional and school catering, are a division of Trust House Forte; there are similar specialist divisions in Grand Metropolitan, e.g. hospitals.

Three points emerge here. The first is the part played by groups having diverse interests in property, hotels and catering, shipping and food processing. Groups, such as those named above, not only straddle the food chain at several points but their strategic direction may lead them to trade one subsidiary activity for that in another sector, and it becomes an intriguing question as to how they can be brought within the orbit of current competition and merger policy, except on the grounds of the amount of capital invested — the present £30 million benchmark.

The second concerns the sources of possible competition and therefore the arena within which the likely performance of rivals is likely to be played out. This is an acknowledgement that groups — other than those that are primarily seen as predators — may come from other than proximate industries, whether within the UK or elsewhere. There is then the requirement, under present UK policy, that there be a presumption that their behaviour on entering an industry or market is going to be harmful to the public interest through its impact on competition. It is easy to appreciate the preference of many of those engaged in monitoring these developments for clearer guidelines, for example, specific practices that are to be deemed prima facie as being against the public interest.

The role of groups of various kinds also presents the problem of the extent to which their divisions compete with one another and the basis of transfer prices between such divisions. How far do the wholesale divisions work to an overall strategy alongside the catering or retail divisions, or how far are they each at 'arms length' from one another? Such questions might have an increased bearing if the process of competition is qualified by concentration and interactive groupings among firms that, whatever their scale of operations outside the chain, have a major role, at least at one stage in that chain.

However, the preoccupation of manufacturers in this and other industries is with the concentration of retailer power. This is seen as having major effects on the choice and design of products, the terms upon which they can be sold, and therefore on the very profitability of manufacturers, even the very largest of them. Economies of scale, while very attractive to both stages, carry risks of vulnerability, in terms of continuity of business where a small number of retail groups can command massive purchasing power in relation not only to the large number of smaller manufacturers who continue to supply them, but even to the large purveyors of branded products who have hitherto had much more punch in the eventual market battle.

There is, therefore, a concern that further retailer concentration should not occur. What is feared is collusion, if not direct collaboration, among very powerful groups. Their impact on processors — or other suppliers — extends to future opportunities for access of branded goods through specialist displays that would have to be manufacturer-funded rather on the shop-within-store approach encountered in department stores. This is over and above the questions of product design and specification already discussed, or the matter of discounts not available to other retailers or even major retail groups.

As long, however, as the competition is both strong and concentrating even more on product variety and quality, then there are opportunities for suppliers. These may favour those who are either smaller but with opportunities for combining for bulk delivery or the outcome of restructuring within larger manufacturing groups.

At the distribution stage there remains one other factor affecting the opportunities for competition, at least locally, namely the planning policies of local authorities and their attitude to the newer types of out-of-town shopping centres. The more these authorities are constrained by public expenditure controls, the more their developments will be dependent on the organisations that can fund new projects and, therefore, the arrangements between such developers and major retail groups. This is evident in the joint initiative involving powerful but not directly competitive groups — food and DIY stores, for example. The linkage of compatible smaller specialist outlets, alongside supermarket projects, is already common. What is interesting to explore is their effect on competition and on the rival attractions of adjacent development points.

But these out-of-town projects leave behind a submerged grouping within society in the inner-city areas where such developments do not materialise and where very little local competition may persist. This too requires further attention.

SUMMARY

Competition, however defined, and whatever the arena taken for study, remains a powerful force for consumer protection, innovation, and balanced development. The impact of competition policy as such varies significantly between the stages in the food chain but is desirable within each one. The problem is to guard against the development of overweening power at any one stage, its compounding by collusion, and protective policies dressed up as 'orderly marketing'.

The structure, both of ownership/control and of activities/products at each stage, is now in a state of flux. It would be equally wrong to try to freeze the power battle at any particular state of play or to assume that no monitoring of the battles is required.

Public and Political Implications

WYN GRANT

Department of Politics, University of Warwick, UK

The shift from a 'politics of production', based around a distribution struggle between management and labour, to a 'politics of consumption', focussing on environmental issues, is outlined. One area in which this shift of focus impacts on the food chain is that of additives. The emergence of the food additives issue is discussed with reference to the role of the media in influencing public attitudes. Concern about additives is particularly marked among the upper social classes and younger age group. New campaigning organisations and coalitions have been formed by 'food activists'. Responses by the food industry to a new political situation are reviewed.

INTRODUCTION

In a recent analysis[1] I argued that the types of political issues facing businesses in the 1990s (not least those engaged in the food chain) were likely to be of a different character from those which they encountered in the 1970s. At the risk of some over simplification, the 1970s were characterised by a struggle over what may be termed 'the politics of production': employer–union conflicts; government incomes and prices policies (the latter particularly important for the food processing industry); and a general distributional struggle over the roles of management and labour in the production process. In the 1980s, these conflicts have receded into the background (although it would be unwise to assume that they have disappeared altogether). In their place, one finds a new series of conflicts, what might be termed conflicts over 'the politics of consumption', in contrast to the former emphasis on 'the politics of production'.

The politics of consumption focusses on environmental issues, ranging from the quality of the air we breathe or the water we drink through to the food we eat. One manifestation of this enhanced environmental concern is the increased legislative representation of 'green' parties in West European legislatures although not, of course, Britain. However, the 'movement' that I am talking about (really a movement of ideas) is broader than the green movement, if one means by that people who are committed to the creation of a society with very different priorities and values from those which have been accepted as the norm in western industrialised societies. Persons involved in the Green Party or organisations such as Greenpeace are a minority (albeit a vocal and highly active one) compared with the nearly three million people estimated to belong to environmental groups in Britain.[2] One must also take account of the large numbers of persons involved in various kinds of consumer organisations, as well as individuals whose occupations lead them to have a professional interest in environmental issues.

I do not have the space here to engage in a lengthy discussion of the origins of this heightened environmental consciousness. Clearly, an important factor has been higher levels of education, leading some segments of the population to demand more information about substances or processes that might affect them, and to take a more critical and sceptical attitude towards those engaged in environmental regulation. The media have played an important role in arousing awareness of environmental issues, although it is difficult to say whether they have followed public opinion, or created it; probably, they have acted as a catalyst. There has also been some 'spillover' effect from the preoccupation with health issues in the United States. Against this general background of concern about environmental questions, it is possible for relatively small numbers of activists to place new issues on the public agenda, as will be apparent later when the controversy over food additives is discussed more fully.

IMPACTS ON THE FOOD CHAIN

Clearly, this increased emphasis on environmental issues impacts on the food chain at a number of points. If the current increase in the number of vegetarians accelerated, there could be a significant impact on the livelihoods of livestock farmers. If major restrictions were placed on the use of pesticides, arable farmers would have to make substantial changes in their methods of production. Of course, changes in production or consumption patterns resulting from environmental pressures need not mean a less viable future for the various participants in the food chain, although clearly there are substantial 'sunk costs' in existing patterns of production. However, farmers can make money out of 'organic' produce, just as the food processing industry can secure good profits from 'additive free' foods.

Rather than look at a range of environmentally driven issues affecting the food chain, which would result in a rather superficial analysis, I have decided to concentrate on one issue which is highly topical, that of food additives. (It also raises issues about nutrition and about labelling which are also topical.) As a consequence, my discussion necessarily focusses on the food processing industry (and on retailers) rather than on farmers. However, major changes of practice in the use of additives would have implications for patterns of farming.

I do not intend to enter here into the controversy about the desirability of additives in general, or about particular additives, or about the concerns that have been raised by writers such as Cannon about the processes by which additives are approved. Rather, I am concerned with why the additives issue has appeared on the political agenda, and what the implications are for the food industry. It is, however, worth noting in passing that there are many products that could not be made without additives; others would have a much shorter shelf life, and might consequentially be a less attractive purchase for consumers. It must therefore be of concern to the food industry that one woman in three believed that there was no need for additives.[3] It must also be of concern that there is 'A serious and widespread conception ... that E numbers generally designate potentially harmful artificial additives'.[3]

WHO CARES ABOUT ADDITIVES?

The survey of consumer attitudes to food additives commissioned by the Ministry of Agriculture, Fisheries and Food[3] provides us with an interesting profile of the type of consumer who is concerned about food additives. Respondents were asked to rank six items which they thought were potentially damaging to their own health. The 'ingredients in food we eat' came third after smoking and environmental pollution.

However, levels of concern about food additives were higher among women, particularly those in the 25–34 age group. There were also some regional differences: 'Respondents in Scotland and Wales appeared less concerned about ingredients in food we eat than the English (8% and 11% respectively) and indeed respondents in Scotland were more concerned about the amount of alcohol drunk'.[3] Women in the south of the country were also more likely to have made purchases of 'additive free' food.[3] Moreover, they were more likely to say that they would be willing to pay 10% more on their shopping bills for additive free foods. This finding was particularly marked for women in Greater London, 85% of whom would definitely or probably be prepared to pay more.[3]

Everything we know about the environmental movement suggests that it draws its strongest support from middle and upper class groupings, and this general conclusion is supported in the particular case of food additives. Respondents in the AB and C1 social classes were more likely to name food additives as damaging to health than those in the C2 and DE groupings, although it should be noted that age differences were an even stronger differentiator on this question (with the 55+ age group taking a much more benign view of food additives than those between 25 and 34). What is perhaps more significant is the way in which social class differences influence reported behaviour. Of AB women, 50% claimed that they usually or always read the list of ingredients on a product label, compared with only 20% of DEs. AB women are around two and a half times as likely to have bought 'additive free' food as DE women.[3]

As well as quantitative research, the study carried out for MAFF had a qualitative dimension, with an analysis of structured discussions with six groups with different age, social class and family circumstances profiles. The report on these discussions notes that:[3]

> Although only a 'consumerist' minority spontaneously referred to the additives, chemicals or pesticides in food as a prime concern, unprompted discussion of those issues was extensive and indications emerged that interest in such topics will continue to grow given continued media interest and coverage.[3]

Some of the individuals in the groups most concerned about additives were young singles (from an idealistic standpoint) and young mothers (concerned about the child health implications).

WHY HAS THE ISSUE EMERGED?

The MAFF study underlines the fact that media publicity has had a major impact on public attitudes towards food additives. Of the women interviewed, 59% were aware of recent publicity on the subject, 65% remembering it as part of a television programme.[3] The qualitative discussions showed that 'Most knowledge was media led and so was highly dependent on issues which happened to be taken up, perhaps solely for their news value'.[3] Respondents could recall a number of items from television programmes such as 'The Food Programme', but 'Women's magazines and newspapers were also influential . . . via their regular coverage of nutrition, diet and additives'.[3] An active minority had purchased the book *E for Additives* which was clearly regarded as something of a bible. The author of this book, Maurice Hanssen, is President of the Health Food Manufacturer's Association of Great Britain and, of the European Federation of Health Product Manufacturer's Associations.

This media interest in the subject has, however, been fostered by a relatively small number of food activists, operating through a number of organisations. Some of the key individuals have published influential books on the subject: *E for Additives* and Cannon's *The Politics of Food*. Another book in this genre, which criticises the scientific basis of toxicological testing and makes the claim that risks are being taken with the health of consumers, is Erik Millstone's *Food Additives*.

At the centre of any map of the new 'cause' groups concerning themselves with food additives stands the London Food Commission (LFC). This body was originally established under the auspices of the GLC, but obtained funding from a trust which outlived the Council's demise. The LFC has operated with a small staff, contracting researchers to prepare reports. These are often well prepared, if one sided, and attract considerable media attention; newspapers often refer to the 'independent' LFC.

Although the LFC has attracted attention as a specialist organisation concerned with food policy questions, the broad spectrum of organisations interested in such issues was reflected in the formation in December 1985 of FACT (Food Additives Campaign Team). FACT was launched with support from Conservative, Labour and Liberal MPs, its stated objective being 'to encourage a healthy food supply in Britain, in particular by a more responsible use of food additives.' An introductory statement issued in June 1986 claimed that:

The widespread use of additives is now a menace to public health. Most additives are the means whereby low quality ingredients, saturated fats and sugars, can be disguised as good nutritious food. Many common additives cause a number of illnesses in vulnerable people, notably children.

It is interesting to examine the character of the 29 organisations belonging to FACT, as it gives some indication of the range of bodies that have concerned themselves with food additives issues. They may be broken down into six categories:

1. Specialist campaigning organisations concerned with food policy issues such as the London Food Commission, the London Hazards Centre (which has funded LFC research), and the Food Additives Petition.
2. Organisations concerned with particular medical conditions or categories of medical condition, or representing paramedical workers (the largest single category, 13 in all): for example: Action Against Allergy, Coronary Prevention Group, Health Visitors' Association, Hyperactive Children's Support Group, National Eczema Society.
3. Unions concerned with the food industry such as the Baker's, Food and Allied Worker's Union and the Transport and General Worker's Union.
4. 'Green' organisations such as Friends of the Earth and the Vegetarian Society.
5. Womens' organisations such as the National Council of Women of Great Britain and the National Federation of Women's Institutes (remember that the MAFF research showed a heightened awareness of the additives issues among women).
6. General campaigning organisations such as the British Society for Social Responsibility in Science and the Campaign for Freedom of Information.

One group of professionals who have taken a close interest in food additives issues include persons who work in paramedical occupations (such as dieticians) or who have a paramedical role in a local authority (such as environmental health). Many of these professionals are involved in a new organisation called the Public Health Alliance which was launched in the summer of 1987 at a conference in Birmingham on 'Rethinking Public Health' and which is operating from the offices of the Health Visitors' Association. Its leader is Dr David Player, formerly director general of the now closed Health Education Council. He has stated that the Public Health Alliance will:

> Seek to expose and make clear the relationships (and ethics) between interests of MPs and others in authority to forces which are antagonistic to health such as the tobacco industry, alcohol industry, and parts of the food and pharmaceutical industries.[4]

None of these groups has very substantial financial resources, but they are linked together by a network of close personal contacts (key individuals occupy leading roles in more than one organisation). Their main impact comes from the access they have to the media, and hence the opportunity to influence both public opinion and decision makers. Assertions are often made on the basis of limited evidence, or contrary evidence is overlooked. For example, FACT claimed that 82 000 people in the UK are allergic to common food additives, a figure which is a rough estimate from an EC report. It also overlooks the fact that half a million people suffer allergic reactions to so-called 'natural' foods.[5]

THE CHALLENGE FACING THE INDUSTRY

Unfortunately, misleading statements made by campaigning groups reach a receptive audience among the public who are suspicious of the good intentions of the food processing industry. The MAFF research showed that 'Attitudinally, manufacturers were now rarely trusted, because they were seen as, or suspected to exert little control and "conscience" over their use of additives'.[3] Retailers were regarded as having taken more positive steps to aid consumers, both by providing information, and by stocking additive free products. Even so, consumers clearly have many misconceptions about additives. Reference has already been made to the widespread belief that 'the E number system identifies ingredients to be avoided'.[3] It is also clear that there is a great deal of folklore circulating about hyperactive children, and that food additives are regularly seized upon as an excuse for bad behaviour. For example, one respondent in the group discussions for the MAFF survey commented, 'I have a neighbour whose little boy was always playing up and their doctor said it was that thing in orange squashes and he's much better now'.

There is clearly a massive task of public education facing the food industry. The MAFF pamphlet, *Food Additives — The Balanced Approach* should help if it is sufficiently widely circulated. As part of its programme of increasing emphasis on the subject of food additives, the Food Sector Group of the Chemical Industries Association has published a booklet *The Chemistry on Your Table* which has had to be reprinted. There is not space here to fully examine the food industry's attitude towards these questions, but the Food and Drink Federation:[6]

> ... does not consider that the food industry should assume *direct* responsibility for the nutritional status of the population. It believes that the role of these industries is to make available the widest possible range of foodstuffs manufactured and labelled in compliance with the law, and in accordance with the principles of good manufacturing practice.

SUMMARY

The food industry is facing an attack on two fronts (although in practice they are often indistinguishable). Some of the individuals who have taken up the issue of food additives would appear to be motivated by a broader concern about what they see as the inadequacies of the capitalist system of production. The food additives issue then becomes one on which manufacturing can be attacked because it appears to be vulnerable. It is vulnerable because there is a wider public concern, particularly among younger and well educated people, about the information that is available on the composition of the food that they eat (including questions of pesticide and antibiotic residues). Ideologues will never be convinced by information that runs counter to their well entrenched views, but the majority of concerned consumers are receptive to rational argument. Hence, in the long run, it may be necessary to improve the quality and quantity of information available to consumers so that they can decide which foods they wish to purchase. What is clear is that issues of nutrition, food additives and labelling will not disappear, but will continue to be persistent issues in the 1990s in response to fundamental changes in public concerns and expectations. Like other industries, the food industry will have to adjust to a new political map with challenges and problems very different from those of the 1970s.

REFERENCES

1. Grant, W. (1987) *Business and Politics in Britain,* Macmillan, London.
2. Lowe, P. and Goyder, J. (1983) *Environmental Groups in Politics,* Allen and Unwin, London.
3. MAFF (1987) *Survey of Consumer Attitudes to Food Additives, Volume 1,* HMSO, London.
4. Rethinking Public Health: the Public Health Alliance (1987) *The Lancet,* July, 228.
5. Fact or fantasy? (1986) *Food Manufacture,* January, 3.
6. Stocker, T. R. (1985) 'Nutrition — the Food Industry's Role', *FDF Bulletin* No. 1, January, 8-16.

Agricultural Policy and its Implications for Food Marketing Functions

SIMON HARRIS

British Sugar plc, Peterborough, UK

ABSTRACT

The paper contrasts the traditional focus of government policy on agriculture with the need to widen it to take in the whole food chain. It describes the effects of the CAP on the food industry's structure, the performance of first and second stage processors and looks at whether the CAP has led to greater concentration than would otherwise have occurred (it concludes not). Finally the paper examines the effects of the CAP on the food industry's political effectiveness and concludes it has weakened it because the split between agricultural raw material processors and food manufacturers has been emphasised.

INTRODUCTION

It is only in very recent years that the Agricultural Economics profession has started to look at the food industry as such, rather than at the more traditional area of agricultural marketing. Even now most of what work is done on the food industry in Britain tends to be done at academic institutions which are not traditionally engaged in agricultural research — institutions such as the Universities of Bradford and Surrey, for example. It is, therefore, a particular pleasure to me, as an agricultural economist, to contribute this paper to a publication devoted to examining the entire food chain in association with the University of Reading where one of the country's leading agricultural faculties is located.

It would perhaps be helpful to disentangle some of the concepts implied in the title of this contribution. Thus I understand 'agricultural policy' to represent the traditional focus of government policy on farm scale problems and the issues of farm income support and market price stabilisation for agricultural commodities. 'Food marketing', however, I conceive of being all that happens to agricultural products once they leave the farm gate to enter the processing and distribution chain on their passage to the ultimate consumer.

THE STRUCTURE OF GOVERNMENT

Although in Britain, the Ministry of Agriculture also includes in its title 'Fisheries and Food', it is not for nothing that the Ministry is known as 'the farmers' friend' in Whitehall. The only time when the British Government has considered policy for the food chain to be of sufficient importance to require a separate Department was during the Second World War when there was a separate Ministry of Food. When food was de-rationed in the early 1950s, the Ministry of Food was merged (in many people's eyes, submerged) with the then Ministry of Agriculture and Fisheries. The merged Ministry of Agriculture, Fisheries and Food (MAFF) has tended to concentrate on the traditional areas of agricultural policy identified above. Although there is a Food Policy Division within MAFF, headed ultimately by one of the three policy Deputy Secretaries, its effect on the direction of government policy, and even MAFF policy, is not readily apparent to the outside observer.

THE NEED FOR A CHANGE IN FOCUS

It is increasingly evident that the traditional focus of government policy in the agri–food area should not only be re-focused, but also re-described. Instead of 'Agricultural' policy, what is needed is a 'Food' policy — directed to the supply of adequate volumes of food, in sufficient variety, meeting appropriate health standards and presented in forms supportive of modern life-styles.

As the food industry is a part of any society's normal manufacturing and distribution activities, it may seem perverse to call for any special government focus in its direction. Nevertheless, there are good reasons why the existing focus of government policy should be switched from the farm to the entire food chain, of which food manufacturers and distributors form the major part.

1. While governments continue to support agriculture, there will continue to be spill-over effects for the rest of the food chain as, according to the Commission about three quarters of Community agricultural production goes through food processors before reaching the consumer.[1] In the case of the CAP these effects are particularly widespread and frequently deleterious for the food industry. In other words, there is a need to direct some government policy effort toward the food industry in order to offset the effects of agricultural policy.

2. If one of the historic justifications for governmental involvement in agriculture is its significance to the economy, then this argument can increasingly be applied to the food industry. As countries develop there is an inexorable tendency for the relative importance of agriculture in total economic activity to decline. Conversely, as more processing is required in the preparation of food before it reaches the consumer, the importance of food manufacturing and distribution activities tends to rise. Taking some (unfortunately) historic data, one can see that by the beginning of the 1980s the 'food industry' contributed more to GDP in the UK, Germany and Belgium that did agriculture. This contrasted with a country such as Greece where food processing was relatively undeveloped (Table 1). An alternative way of viewing the same phenomenon is represented by calculations of the food industry's share of the consumers' dollar/pound. In the UK, for example, the cost of processing and distribution as a proportion of retail food prices rose from 48·1% in 1973 to 53·7% in 1984[4] and the farmers' share fell conversely. Mordue[5] concluded that agriculture, food manufacturing and food distribution each contributed some 2·5% to UK GDP at the end of the 1970s. The European Commission estimated that although food processing accounted for about 3·4% of Community GDP (almost the same as agriculture), it accounted for 10% of net value added in all industry, making it 'the Community's largest single industry'.[1]

3. Not only is the world market outlook for manufactured food exports better than that for exports of bulk, unprocessed agricultural commodities, but also the value added is higher. As a consequence, governmental policy needs to be concentrated more

TABLE 1
Relative Contribution of Agriculture and the Food Industry to the Economy (1979)

	Germany	France	Italy	Netherlands	Belgium	UK	Greece
Value added (millions, national currency)							
Agriculture	31 820	120 406	18 426	11 423	71 162	4 242	210 000
	(35·1)	(53·4)	(63·3)	(53·6)	(37·4)	(34·8)	(82·0)
Food industry	58 841	105 073	10 679	9 877	119 195	7 948	46 000
	(64·9)	(46·6)	(36·7)	(46·4)	(62·6)	(65·2)	(18·0)
Total agri-food	90 661	225 479	29 105	21 300	190 357	12 190	256 000
	(100)	(100)	(100)	(100)	(100)	(100)	(100)
Employment (thousand)							
Agriculture	1 542	1 890	2 840	279	124	631	1 020
	(63·7)	(76·6)	(85·8)	(62·4)	(51·7)	(51·6)	(94·7)
Food industry	879	576	469	168	116	593	57
	(36·3)	(23·4)	(14·2)	(37·6)	(48·3)	(48·4)	(5·3)
Total agri-food	2 421	2 466	3 309	447	240	1 224	1 077
	(100)	(100)	(100)	(100)	(100)	(100)	(100)

	1979 EC '9' Share of GDP (%)	*1980 EC '9'* No. employed (million)	*1981 EC '10'* Trade balance (billion ecu)
Agriculture	3·7	7·7	−14·4
Food industry	3·4	2·7	+5·0
All industry	29·2	40·1	+12·8

Sources: Commission of the European Communities.[1-3]

on encouraging the export of processed products and less on the disposal of primary commodity surpluses by dumping them on world markets.

4. As foodstuffs become ever more processed to meet the demand for 'convenience' foods, the need for governmental intervention to ensure their quality rises. Advances in food processing technology have been a major factor in making possible a whole social revolution — with the proportion of women working outside the home continuing to rise (therefore widening the market for ready-prepared meals) and the US habit of 'snacking' (rather than sitting down to formal meals) spreading. As foodstuffs become ever more removed from the relatively unprocessed state of the past, consumers become more dependent on the food industry to maintain the health status of their products. Hence, what the food industry does with food is an important and legitimate area for increasing governmental policy attention.

5. With the massive increase in agricultural trade tensions, there is the need for government to become more involved in maintaining, and raising, the food industry's ability to trade with Third Countries. It should be recognised that when agricultural trade disputes flare, it is normally processed foodstuffs which are on the retaliation lists, rather than basic agricultural commodities.[6] Despite its importance, this paper does not go further into this issue as it will be covered fully in a separate paper.

SOME GENERAL EFFECTS FOR THE FOOD INDUSTRY

The effect of agricultural policies for the food industry is, of course, a complex balance of factors. As a generalisation, the closer to the farm gate the more significant the effects of agricultural policy are likely to be. Conversely, the further away from the farm gate the more attenuated the effects become.

How significant the effects of any particular agricultural policy will be for the food industry depends on both the level of support provided to agriculture and the type of policy measures used to provide this support. British agricultural policy pre- and post-EC accession strikingly demonstrates this point. The post-war agricultural support policy followed in the UK, up to EC accession, was based on a system of direct-income payments (known as Deficiency Payments) to bring up returns on agricultural product sales to stipulated Guaranteed Prices. So far as the food industry was concerned, it was free to buy its agricultural raw materials at world prices from wherever it wished and it did not have any involvement in the process of providing support to agriculture. Indeed it is likely that most British food companies prior to EC accession could not have said what form of agricultural support policy was followed in the UK.

Since EC accession the CAP has applied in the UK — as elsewhere in the Community. Its results for the food industry are very nearly the exact opposite of the UK's pre-accession system of agricultural support. Thus the food industry's ability to source its agricultural raw materials from wherever it wishes no longer exists. It is constrained to using Community-produced raw materials for the overwhelming majority of its requirements because of 'Community Preference' expressed through the variable import levy mechanism, in particular, and the system of minimum import prices in general. Again, it can no longer buy its raw materials at world prices because the principal mechanism for agricultural support under the CAP is a system of managed product markets, whereby internal EC market prices are kept at fixed levels — generally well above those prevailing on world markets.

Finally, a system of managed markets is highly dirigiste requiring a mass of widespread and highly detailed legislation which is, to a large extent, implemented through the food industry. As a result the food industry has passed from a state of blissful ignorance about agricultural support pre-EC entry, to one where the CAP features as the food industry's leading *bête noire* and where all major food companies have had to take on specialists in EC legislation just to cope with the flood of Regulations, Directives, proposals, etc. issuing daily from Brussels!

To add insult to injury, agricultural product markets are not managed in order to satisfy food industry desires. Very much the reverse, as the European Commission's prime objective in their management is to achieve the Community's farm-price goals at minimum cost to the Community Budget. Food industry concerns do not feature in the list of priorities. This is reflected in the composition of the Commission's Agricultural Management Committees, where only national civil servants are present. Their concerns are with specific agricultural product markets, but entirely with an agricultural focus so that, for example, their concerns do not normally even extend to ensuring the levels of agricultural product quality needed by the food industry.

EFFECTS FOR FIRST-STAGE PROCESSORS

A broad definition of first-stage processing is that it covers the conversion of farm-gate products to storable and usable ingredients for the rest of the food industry. Thus sectors covered include slaughter houses, flour mills, creameries and sugar factories/refineries. (For a fuller discussion of the difference between first- and second-stage processors, see Harris, Swinbank and Wilkinson.[7])

For first-stage processors, in particular, the effects of agricultural policy are a mix of gains and losses, although most observers would agree, that on balance first-stage processors have benefited from the CAP. Thus in order to provide an open-ended (until recently) support buying commitment as the floor-price mechanism for agriculture, the CAP has provided a guaranteed market for first-stage processors. This results from the fact that many farm-gate products are perishable and, by definition, not readily storable. Consequently, intervention buying applies in many cases to the products of the first-stage processors (e.g. butter and SMP, white and raw sugar, carcass meat) which can be stored.

Again first-stage processors benefit from the high levels of protection under the CAP. Otherwise they would be unable to buy farm produce at the institutionally determined CAP minimum prices. For some processors the situation goes even further as their processing margins, either explicitly (sugar processors) or implicitly (creameries) are set and guaranteed by the CAP.

Maybe most importantly, however, the result of the CAP has been to lead to a major expansion in the volume of the Community's agricultural production. This growth in volume has meant, in turn, more produce for the first-stage processors to handle and hence the opportunity to expand on the back of the remarkable growth in EC agriculture.

With guaranteed and protected markets, growing volumes of raw materials to be handled and continuity of agricultural policy, first-stage processors have been given every incentive to modernise and expand. The result has been to lead to a growth in the volume of capital committed and an expansion in the scale of the industries involved as first-stage processors have taken advantage of the opportunities offered.

The down-side of the CAP for first-stage processors, however, is first their loss of commercial freedom. They are not free to negotiate prices with farmers in a way that they certainly would in a non-managed system. They are not free to purchase from Third Countries and their ability to export to world markets is utterly dependent on the Community setting adequate levels of export refunds.

Second, they are extremely vulnerable to changes in the direction of the CAP, as these may lead to significant changes in the managed market system in which they operate. More seriously, they may lead to sudden alterations in the level of agricultural production the Community is prepared to support. Unfortunately, such changes tend to be abrupt and sprung on the world without warning, as with the imposition of dairy quotas in 1985 which was unheralded, but led directly to the closure of several creameries in the UK. Because of the priority given to agricultural objectives in running the CAP, the food industry has little or no say in changes in the direction of the CAP, even when the food industry may be drastically affected by the changes. Furthermore, compensation from the Community is not generally available for processors caught by such sudden shifts in the direction of the CAP and who have to close capacity as a result.

Third, first-stage processors are the level of the food industry which gets caught, in particular, by the detailed governmental regulation associated with the CAP. This effect has been pointed to already. It is a serious factor, however, when management is trying to operate commercially and yet, at every turn, is hedged about with CAP-derived restrictions.

EFFECTS FOR SECOND-STAGE PROCESSORS

These are more normally termed 'Food Manufacturers'. They take processed raw materials and produce manufactured food items from them. They produce branded products with increasingly sophisticated and diversified presentation and packaging and they 'practise all the techniques of non-price competition in marketing e.g. advertising and sales promotion'.[8]

Despite their distance from the farm gate, most observers would accept that, on balance, food manufacturers are worse off as a result of the CAP. At the most basic level, they are caught between the CAP's results in raising market prices for agriculturally-based products and the final consumption market, where it is often not possible just to simply pass through raw material price increases. They are faced with selling to retailers in particular, who have highly concentrated market power to squeeze out the most favourable terms from manufacturers. In any case although demand for food as a whole is inelastic, many food items are significantly price elastic (particularly those containing large elements of processing) and the ability to pass on price increases resulting from the CAP is circumscribed.

Again, as with first-stage processors, food manufacturers cannot buy raw materials freely from Third Countries. As a result, they are forced back on using domestically-produced agricultural products and cannot avoid the impact of the CAP in raising market prices for farm products. This system reaches the heights of absurdity in those instances where the Community either does not produce specific food ingredients at all, or does not produce the required types or qualities. For items such as high diastatic barley, 'strong' wheats, some types of maize, etc. the impact of the Community's agricultural import provisions is merely to tax food manufacturers and, ultimately, consumers, without benefiting any domestic producers, as such items still have to be imported from Third Countries.

Unlike first-stage processors, however, food manufacturers do not have guaranteed markets for their products nor the possibility of institutionally-determined processing margins. This puts food manufacturers, therefore, into a fundamentally more competitive world than first-stage processors. This is not to say that first-stage processors do not operate in a competitive environment but merely that the effects of the CAP are to insulate them from *some* of the normal commercial pressures. Although, undoubtedly, many first-stage processors would feel that they are in an extremely competitive environment with the pressures on for rationalisation and modernisation leading to mergers and factory closures. Implicitly, the CAP provides some cushion in terms of slowing the pace of structural adjustment, to which all industries are subject, but not ultimately preventing it. But for food manufacturers, the CAP does not provide any cushion at all.

In one respect, however, the CAP alters conditions of competition amongst food manufacturers. This is in the area of food ingredient purchasing where the CAP's results in fixing minimum market prices mean that it is not possible to obtain significant competitive advantage by purchasing more keenly than a competitor. For all companies there is little variation in CAP food ingredient prices, apart from the normal range associated with quantity discounts and, maybe, geographical location.

A specific feature of the CAP which affects food manufacturers is the distortions it causes between food ingredient prices. The price relationships of food ingredients in the CAP are not the same as those which would apply under free market conditions. As a result, food manufacturers are encouraged to adopt product formulations and to seek alternatives to CAP commodities which otherwise they would not need to.

In one very major respect, however, food manufacturers and first-stage food processors share the same concerns — their competitiveness on world markets is dependent on the European Commission providing adequate levels of export refunds. These are necessary to compensate for the higher levels of CAP raw materials, as compared with world prices. In the food industry's view, export refunds are not export subsidies, but merely compensation for inflated CAP prices.

EFFECTS OF THE CAP ON FOOD INDUSTRY STRUCTURE

To assess the effects of the CAP on the structure of the food industry is difficult, not only due to the lack of data on food industry structure as such, but also because it would imply hypothesising what the food industry's structure would have been in the absence of the CAP. All that is possible is some impressionistic comments on what appear to be effects resulting from the CAP.

In general, the size of the first-stage processing sector is probably larger and financially stronger than it otherwise would have been. The implications appear to be that first-stage processing companies form a suitable base for expansion in the rest of the food industry. Perhaps the most striking example of such an expansion is that of the Ferruzzi Group in recent years, not only expanding in the food industry but also into the chemical industry.

This is not to say that food manufacturers have not been able to continue to grow and concentrate their structure among a few large companies which have then been able to vertically integrate up and down the food chain. Presumably, insofar as first-stage processors are financially stronger than they would have been, this may have had some influence in reducing the ability of food manufacturers to integrate backwards. In practice, of course, national differences in industrial structure and stage of development of the food chain, as a whole, tend to mask these presumed effects of the CAP.

It must be a moot point whether the CAP has had any effect in terms of increasing the concentration in the food industry beyond that which would have happened in any case. The trend to increasing concentration among, in particular, food manufacturers and retailers is well known and long established, as is the continuing decline in the importance of wholesalers. It would seem feasible to hypothesise that some of the concentration seen among food manufacturers, in particular, is a response to being caught between the pressures exerted by the CAP on raw material prices on one hand, and that exerted by the retailers on the other hand. But categoric statements do not seem possible about the role of the CAP in encouraging, or otherwise, greater concentration in the food industry. Watts[9] has pointed out that in 1979, of the UK's ten largest food companies only half of their sales were food — as they had become increasingly diversified.

Two features of Community food industry structure are, however, more easily tied to the operations of the CAP. First, there is the general encouragement for agricultural co-operatives — seen as being desirable by the Community. The result is that, on the Continent in particular, agricultural co-operatives own a significant proportion of first-stage processing capacity. Of course, the co-operative movement in agriculture appears to be more firmly entrenched on the Continent than in Britain in any case. It is interesting that in a very different environment (the USA), agricultural co-operatives have taken important stakes in grain exporting and meat processing. So again the same point is raised — how far would the significance of agricultural co-operatives have developed without the CAP?

Second, CAP influence on the food industry means importing is discouraged. This was most graphically illustrated in 1974–7 when the Community introduced a ban on the import of meat, without warning. As a result, in one fell swoop, most of the companies involved in the import of meat in Britain had to be closed down virtually overnight. True to mercantilist principle, importing is discouraged but exporting encouraged. The consequence is that the Community's food industries are heavily geared toward exporting to world markets, but not to importing from them.

A disappointing feature of the Common Market so far has been the lack of integration between industries in the different Member States. The only food companies which have taken advantage of the Common Market's existence to design Community-wide strategies for manufacturing and

TABLE 2
The World's Top Thirty Largest Food and Drink Companies, 1986/87[10]

	Country of incorporation	Sales (million US dollars)	Profit (million US dollars)	Return on capital (%)
1. Unilever	The Netherlands/UK	27 129	1 279	22·3
2. Nestlé	Switzerland	23 626	1 225	12·8
3. Philip Morris	USA	22 279	1 842	27·0
4. RJR Nabisco	USA	15 868	1 209	20·0
5. Pepsico	USA	11 500	595	23·7
6. Kraft	USA	11 011	489	25·8
7. Sara Lee	USA	9 155	267	18·9
8. Conagra	USA	9 002	149	20·6
9. Beatrice	USA	8 926	N.A.	N.A.
10. Hanson Trust	UK	8 890	891	29·2
11. Anheuser-Busch	USA	8 258	615	21·3
12. Grand Met	UK	7 742	719	25·5

13.	Coca-Cola	USA	7 658	916	28·4
14.	Dalgety	USA	7 631	118	21·6
15.	Barlow Rand	South Africa	7 618	221	20·2
16.	Borden	USA	6 514	267	16·1
17.	BSN	France	6 181	258	9·9
18.	Elders IXL	Australia	6 169	232	10·5
19.	Taiyo Fisheries	Japan	6 162	1	0·4
20.	Pillsbury	USA	6 128	182	13·2
21.	Ralston Purina	USA	5 868	523	54·4
22.	Archer Daniels M	USA	5 775	265	11·2
23.	Snow Brand M P	Japan	5 390	24	5·5
24.	General Mills	USA	5 208	222	30·4
25.	Hillsdown	UK	4 981	145	20·1
26.	CPC Int	USA	4 903	355	32·7
27.	Heinz	USA	4 639	339	24·3
28.	I C Industries	USA	4 581	252	16·0
29.	Allied-Lyons	UK	4 549	286	10·8
30.	Quaker Oats	USA	4 539	244	22·4

marketing have been the multi-nationals such as Nestlé, Nabisco, General Foods, Kellogg and Heinz. Even then strategies are generally linked to a series of national markets within the Community, rather than a single unified market. Although the development of the Ferruzzi Group has been mentioned above, it is difficult to think of other EC-owned food industry companies which have been able to spread their national base to develop markets and, if necessary, production in all the other Member States. One major reason why EC companies have been less successful in geographical diversification than the multi-nationals is, of course, their difference of scale (Table 2) with only five EC companies featuring in the world's top 30 food companies.

Although the Community's development of the internal market, with its hope of a single, unified market in 1992, would appear to have little to do with the CAP, nevertheless barriers provided by agricultural and food policies seem likely to form some of the most intractable barriers to sweeping away national frontiers. In particular, the presence of mcas is likely to continue to form a barrier to paperless frontier crossings. Although the EC is meant to be sweeping away mcas by 1992, it is difficult to see how this is to be achieved in practice, unless a much greater degree of convergence between Member State economies is possible than has been the case up to now. It would be ironic and deeply disappointing if the Community's first fully developed common policy, in the shape of the CAP, were to prove a barrier to the completion of the internal market programme so necessary to forming a single, integrated market across the entire Community.

THE CAP'S EFFECTS ON THE POLITICS OF THE FOOD INDUSTRY

An under-rated effect of the CAP, in political terms, is its creation of an agri–food complex whose interests are largely common. Most first-stage processors are able to develop common policy approaches with farmers and so make a more resistant bloc to change. Broadly, the political divide between those wishing to raise CAP support prices and those opposed to price rises generally falls between the first- and second-stage processors, with the first-stage processors tending to line up with the farmers.

Food manufacturers tend to see themselves as aligning with

consumer interests in having a shared desire to hold down food prices. Nevertheless, this potential grouping seems, in practice, to carry little weight. The more successful pressures in holding down CAP prices have come from the Community's budgetary difficulties on the one hand, and the difficulty of disposing of farm surpluses, on the other hand, when world markets for soft commodities are, as a generalisation, over-supplied. Neither consumer, nor food manufacturer groupings have been major influences in holding down CAP prices, in ecu terms, in recent years.

Josling and Ritson[11] have commented that 'perhaps the greatest enigma in the politics of food is the lack of overt influence of the food industry *per se*'. It may be that one of the most significant effects of the CAP on the food industry is emphasising the split between first- and second-stage processors. This in turn is a major factor in the food industry's inability to come to a united view on many issues concerned with the CAP.

It is perhaps not surprising that the food industry's European level trade association (the CIAA) has, as a consequence of this split in particular, great difficulty in reaching unified viewpoints. The CIAA's problems are compounded by the influence of agricultural co-operatives on the delegations from some countries, because of their ownership of processing facilities. There are many other factors contributing to the CIAA's ineffectiveness, but it is perfectly clear that the CAP is not a factor pushing the food industry to greater unity and a more effective voice in the discussion of policy. The CAP is rather a factor contributing to the food industry's disunity and lack of effectiveness in influencing policy.

Other factors leading to policymakers' ability to ignore food industry concerns include the fact that the food industry is not seen as a united whole, but as a series of individual sectors each with its own preoccupations. This contrasts with agriculture which is perceived as a single industry despite the major differences it encompasses.

Again, support for agriculture is institutionalised in a way that it is not for the rest of the food chain. As Gray[12] pointed out the Treaty of Rome provides for support for agriculture, but 'there are no articles which provide for industrial policies for the downstream industries'. The consequence is that the political influence of agriculture is entrenched, whereas the food industry's ability to exert political influence on the CAP is weak.

CONCLUSIONS

Despite a glaring lack of information, it does seem there are several conclusions that may be drawn.

1. The focus of government policy should be switched from its concentration on agriculture to a wider view of the food chain as a whole.
2. The effects of the CAP for the food industry are the reverse of the UK's pre-accession system of agricultural support.
3. As a generalisation, first-stage processors have gained from the CAP, while food manufacturers have been disadvantaged by the CAP.
4. It is not clear whether the CAP has led to greater concentration in the food industry than otherwise would have been the case.
5. One of the reasons for the food industry's ineffectiveness in political terms may be the CAP's influence in emphasising the difference between the first- and second-stage processors.

REFERENCES

1. Commission of the European Communities (1983) *The Agricultural Situation in the Community 1982 Report,* Brussels, pp. 27, 28.
2. Commission of the European Communities (1982) *The Agricultural Situation in the Community 1981 Report,* Brussels, p. 281.
3. Commission of the European Communities (1984) *The Agricultural Situation in the Community 1983 Report,* Brussels, p. 49.
4. Harvey, D. (1987) *The Future of the Agricultural & Food System,* EPARD Working Paper No 1, University of Reading, p. 133.
5. Mordue, R. (1983) The Food Sector in the context of the European Economy. In: *The Food Industry,* J. Burns, J. McInerney and A. Swinbank (Eds), Heinemann, London, p. 40.
6. Harris, S. (1988) *The British Food Industry's View of the Uruguay Round of GATT Negotiations,* FDF, London, p. 3.
7. Harris, S., Swinbank, A. and Wilkinson, G. (1983) *The Food and Farm Policies of the European Community,* John Wiley, Chichester.
8. OECD (1983) *OECD Food Industries in the 1980's,* Paris, p. 13.
9. Watts, B. (1983) Structural adjustment in the UK food manufacturing industry over the past twenty years. In: *OECD Food Industries in the 1980's,* OECD, Paris, p. 118
10. Bertele, U. (1988) Creazione del Mercato Unico Europ, *Materie Prime,* **3,** September, Nomisma, Bologna.

11. Josling, T. and Ritson, C. (1986) Food and the Nation. In: *The Food Consumer*, C. Ritson, L. Gofton and J. McKenzie (Eds), John Wiley, Chichester, p. 17.
12. Gray, P. (1984) A Commission view of the Food and Drink Industries. In: *The EEC and the Food Industries*, Food Economic Study No 1, A. Swinbank and J. Burns (Eds), University of Reading, p. 146.

Report of Discussion

Rapporteur: JIM BURNS

Department of Agricultural Economics and Management, University of Reading, UK

The discussion raised the fundamental issue of the space within which public policy takes place, and the associated problem of defining the market or industry to which policy action should refer. The food sector is becoming increasingly EEC and internationally orientated, and while for many policy purposes the EEC dimension is paramount, it is still evident that aspects of, for example, competition policy are largely UK focussed. However, the movement towards completing the internal market in 1992 is likely to enhance the Community-wide basis of policy, although it may be that less policy and more market forces are conducive to a freer trading environment inside the EEC.

As this publication emphasises, policies should take into account the whole food chain rather than simply a particular link. This approach raises questions as to whether policy institutions are sufficiently concerned with the implications of measures for others in the chain. At an EEC level, for example, one directorate-general of the Commission, DG VI, is concerned solely with agriculture. In the UK, the MAFF has an obviously wider remit, but questions of balance still arise. A related basic issue is that many policies not directed specifically at food have importance for those in the food chain. These include taxation and monetary policy, foreign policy, health and environmental matters. It is not always apparent that the policy-making process has considered these interrelations.

At several points in the discussion, the significance of the Common Agricultural Policy was highlighted for members of the food chain beyond the farm gate: processors, manufacturers, traders, retailers and caterers. It was felt that, in pursuing farmers' interests, policy-makers

307

have at times adopted a restricted approach which limits the availability of alternative agricultural raw materials to manufacturers, and at others taken a blanket approach which pays little heed to the detailed implications of the regulations for users. Administrative procedures which replace market forces may not act in the best trading interests of food manufacturers, many of which, it was suggested, would prefer a more liberal approach to trade. While EEC Authorities may see marginal adjustments to policies as significant, others would prefer a greater involvement of agriculture in the international tariff discussions under GATT, and in general a closer linking of agricultural and food trade issues. Some, indeed, saw the end of the CAP as a desirable goal, but others considered that even more highly protected agricultures elsewhere would need liberalising at the same time. One particular aspect of the CAP singled out as not working well was the export refund system. The complexity of, and lack of transparency in, the calculations was considered to be an obstacle to trade. However, it was noted that price considerations are by no means the only factor in trade, especially with respect to more highly processed foods.

Consumers are also less concerned with the price of foods than before, as variety, quality and health matters assume greater importance while the major influences on consumers' behaviour are not always well understood, the role of the media is thought to be significant. Reports of scientific matters have not always been accurate, and some argued that consumer misconceptions over, say, health risks in food had been exacerbated by media coverage.

Views differed as to the likely impact of the completion of the EEC internal market in 1992. Partly it would depend on how governments reacted in terms of their other policies and approaches: particularly in national competition policy, control of media, labour and population mobility. In practice, many manufacturers already are moving towards a European or international orientation, and while 1992 remains important, other factors, too, would influence investment decisions.

All recognised the importance of market — as opposed to production — orientation in the food chain. This implies greater consumer choice, market competition and transparency. Policies need to be recognised for what they are. The CAP, for example, is principally a social policy which operates by managing markets rather than a market policy as such. However, in joining the EEC the UK accepted the CAP and other established policies, and policy developments necessarily now take place as a series of compromises between the different interests of the 12 EEC nations. Conflicts also arise between the interests of members of the same industry, between industries and between policy-makers of differing philosophies.

Summary

C. R. W. SPEDDING

Department of Agriculture, University of Reading, UK

There is little point in attempting to summarise the contributions, literally, over such a wide range of subjects. But there is value in participants being offered a view of what has crystallised in the mind of one of the organisers. This supplies conclusions against which those of each participant can be tested, to determine, by agreement or disagreement, what one's own views are.

Although a number of people clearly felt that the time had come to move the emphasis in food production from quantity to quality, it is well to remember that this is a UK (or developed country) view that does not hold worldwide.

Most participants recognised the need to think about the food chain as a whole, although some felt that, because of the complexity of the interconnections, a food net might be a better description. I retain the view that, although the context and environment are important, the purposeful nature of the activity, directed as it is to meeting the needs of consumers, makes the food chain a more relevant image.

However, the conference made it clear that none of us had a very satisfactory picture or model of the food chain and that it would be worth some effort — probably on the part of an interdisciplinary group — to rethink and elaborate a model. Adding more detail, such as input industries and waste products, could be done relatively easily.

But the sheer size and importance of the catering sector, which surprised many of us, and the impact beyond consumption on the health of the consumer are examples of whole new dimensions that need to be considered.

The discussions gave proper weight to the importance of the consumer by starting with public perception and its inadequacies. This

311

revealed the need for recognised professional status and qualifications for nutritionists and the need for education in schools and colleges.

It is clear that different people mean different things by education, especially in relation to the public, varying from the provision of clear and accurate information to the persuasion of others to one's own point of view.

Everyone regrets misinformation flow and favours jargon-free communication: most favour the scientist getting involved in public debate, provided the message is clear. Unfortunately, scientists have to cope with genuine scientific uncertainties and the possibility that new evidence will alter the message. In addition, the media are not always seeking balance and the listener may be already biased.

But bias may be reasonable, in the sense that ordinary people, including schoolchildren, already attach perfectly legitimate meanings to words that specialists use in a quite different way. This is not just a matter of jargon or ignorance. Words like 'energy' or even 'life' are examples of this.

Clearly the communication system is far from perfect but we have to use it. However, trust is vital: no matter how clear communication is, the result will be unsatisfactory if the spokesman (scientist, nutritionist, etc.) is not trusted.

Even if we believe that we know what is right, in relation to a healthy diet, there are many different views as to the extent to which such things should be subject to planning, control, advice, persuasion or just information provision.

The answer that finds most favour involves a combination of information and choice. Informed choice is, in fact, the only way of determining what and how much is actually wanted.

Even in university education it may be desirable to increase the choice of what is studied and how much vocational element there should be in courses.

Vocational training is not simply a matter of techniques, however; it should also involve, for example, encouraging teamwork and preparing people to engage in it, as they will have to do later in their careers. Where teamwork is already part of the educational process, it is worth asking, where and how is it rewarded?

The main problems of relating education to the importance of the food chain remain: how is the student of agriculture or of food science to be made aware of the food chain, of which his subject is only a part?

A similar problem occurs in relation to the establishment of priorities for R & D. How can the complexity of the whole food chain be taken into account in doing this? Very little attempt has been made and it appears very important that this should be put right.

Whatever part of the food chain is worked on, or proposed to be worked on, it is clearly vital that someone should be able to work out the total consequences of resulting changes on all parts of the food chain and not merely on the sector under investigation. The capacity to do this barely exists, so it is hardly surprising that this job is not being done.

Breadmaking supplies a very good example — in the Chorleywood process — of how R & D in crop production may be outdated by technological developments in processing. The advantages of the process to the UK are fairly clear but the impact on other countries, such as Canada, are not always so obvious. This reminds us that some food chains are international, often involving developing countries.

Perhaps relating R & D priorities to the whole food chain would be facilitated by the, in any case desirable, greater dialogue between research workers and industry that is clearly required. But the acceptability of new technology to the consumer should also be taken into account in assessing priorities in R & D.

The consumer is concerned with technology throughout the food chain — as much in production as processing — and even requires labelling to relate to all parts of it. The customer will often wish to know from a label *how* a product was produced as well as what is in it. Public opinion is one of the influences on public policy and, as a result, policy can change quite rapidly, embracing the environmental context of the food chain as well as the chain itself.

There is a case for food policies, though it is less clear what would be involved in a food policy. However, there is also, because of the existence of the food chain, a case for a *food and agriculture* policy, giving full recognition to the internationalisation of food and the fact that we are increasingly only a component of the EEC.

No conference normally concludes a debate and the hope for this one is quite the reverse. The intention is to *start* one. Certainly there was a general agreement that the conference was a good idea and that the subject is important. If it is to develop further, a great deal of further debate will be necessary in order to crystallise a clear picture of the food chain from the complexities in which it is embedded.

Index

p. 260, 262, 269
294
298